DECODING
THE MESSAGE
OF THE PULSARS

Other books by Paul A. LaViolette

Earth Under Fire: Humanity's Survival of the Ice Age

Genesis of the Cosmos: The Ancient Science of Continuous Creation

Subquantum Kinetics: A Systems Approach to Physics and Cosmology

DECODING THE MESSAGE OF THE PULSARS

Intelligent Communication from the Galaxy

Paul A. LaViolette, Ph.D.

Bear & Company
Rochester, Vermont

Bear & Company
One Park Street
Rochester, Vermont 05767
www.BearandCompanyBooks.com

Bear & Company is a division of Inner Traditions International

Library of Congress Cataloging-in-Publication Data

LaViolette, Paul A.
  [Talk of the galaxy]
  Decoding the message of the pulsars : intelligent communication from the
galaxy / Paul A. LaViolette.
     p. cm.
  Originally published: The talk of the galaxy. Alexandria, VA : Starlane
Publications, ©2000.
  Includes bibliographical references and index.
  ISBN 1-59143-062-3 (pbk.)
  1. Life on other planets. 2. Interstellar communication. 3. Radio astronomy.
  I. Title.

QB54.L38 2006
576.8'39—dc22

2006000293

Printed and bound in the United States by Lake Book Manufacturing

10 9 8 7 6 5 4 3 2 1

Text design and layout by Rachel Goldenberg
This book was typeset in Sabon, with Charlemagne and Avenir as the display
typefaces

Diagrams are by Paul A. LaViolette unless otherwise noted

To send correspondence to the author of this book, mail a first-class letter
to the author c/o Inner Traditions • Bear & Company, One Park Street,
Rochester, VT 05767, and we will forward the communication.

CONTENTS

PREFACE

For decades SETI astronomers have been searching the sky for radio signals of extraterrestrial intelligence origin but have found nothing. Perhaps the problem is that they are looking for the wrong type of signal. They have been seeking discrete frequency transmissions similar to terrestrial AM or FM radio signals. But there is no guarantee that another civilization would be using this particular method of communication. Broadband transmissions, covering the entire radio frequency spectrum, would be a more logical choice because they would be more easily detected regardless of which frequency one's radio telescope happened to be tuned to. Such broadband emission, called "synchrotron radiation," is readily produced by magnetically decelerating a beam of cosmic ray electrons. By arranging that the electrons track in a straight line as they decelerate, the synchrotron radiation can be confined to a narrow beam that has minimal decrease of its intensity over interstellar distances, thereby ensuring that the target civilization will be receiving a strong signal.

This kind of radio transmission essentially describes the signals that astronomers routinely observe coming from pulsars. In particular, there is considerable evidence to suggest that these signals are artificial. Indeed, they are the most complex ordered phenomenon known to astronomy, and to this date, they have not been adequately accounted for by any natural-cause model. The neutron star lighthouse model, for example, falls far short of this challenge. Many astronomers, though, will experience difficulty relinquishing the paradigm they have come to accept, even when confronted with its shortcomings.

In reading this book, keep in mind that several sets of relationships must be taken into account to fully understand the symbolic message that the pulsar network is conveying. One part ties in with another to form a complete picture. Thus, it is helpful to contemplate these findings as a whole. It is also useful to read the books *Earth Under Fire* and *Genesis of the Cosmos,* as they provide background material essential to understanding the pulsar message.

ACKNOWLEDGMENTS

I would like to thank my father, Fred, and also Tom Valone for the long hours they both spent helping me edit this manuscript. I would also like to thank Joscelyn Godwin; Jackie Panting; my sister, Mary; and mother, Irene, for their helpful comments on the manuscript, and Geri Davisson for her help as well.

ONE

THE PULSAR ENIGMA

One other thing it can be is an intelligent civilization attempting to communicate with other worlds, because everybody has said that's how you'd mark yourself. You do something that can't be done in nature. You make the pulse rate of a nearby pulsar exactly right, not deviating in the least year after year.

FRANK DRAKE, 1974

Discovery

It was July 1967. The world's first scintillation radio telescope had just been completed, a device that would allow astronomers to detect rapidly varying radio emissions coming from distant stars. Cambridge University graduate student Jocelyn Bell and her astronomy professor, Anthony Hewish, were making final adjustments to the field of radio aerials that lay stretched out across the English countryside. Little did they know that within a month Jocelyn would stumble upon one of the most important astronomical discoveries of the century. They had finished scanning an area of the sky located in the direction of the constellation of Vulpecula. Jocelyn was looking over the yards and yards of pen chart data that scrawled the signals from their antenna array and noticed something quite unusual. One of the radio sources whose radio signal twinkling they had been observing appeared to be emitting a steady series of radio pulses, or "beeps," each lasting a few hundredths of a second. Hewish at first dismissed the pulses as radio interference from a terrestrial source such as sparking from a passing automobile.

The signal had faded and could not be detected on subsequent observations, but one night it reappeared. After several months of observation, seeing that the signal came from a fixed location in the sky, he became convinced that they had detected a new kind of astronomical source.

At the end of November, after obtaining a suitable fast-response chart recorder, they were able for the first time to get an accurate fix on the timing of the pulses. Six hours of observations had shown that the pulses had a very regular recurrence period of 1.33733±0.00001 seconds. Additional months of observation added two more decimal places to the precision of the source, and today we know its period to better than six parts per trillion, yielding a pulsation period of precisely 1.337301192269±0.000000000006 seconds per cycle! This discovery caused quite a stir among the project's scientists. Nothing like it had ever been seen before. It seemed to them that they may have detected signals being sent from an alien civilization. Months of careful observation had revealed that the radio source lay about two thousand light-years away. The idea that the object was a radio beacon operated by extraterrestrial intelligence (ETI) was seriously considered, for this was the first time in the history of astronomy that a source of such precise regularity had been encountered. In fact, they initially named this source LGM-1, the acronym LGM standing for Little Green Men.[1]

Near the end of December, Jocelyn discovered a second pulsating radio source in the constellation of Hydra, which lay in an opposite part of the sky. This one, which had a period of 1.2737635 seconds, was later christened LGM-2. With the discovery of this second source, the Cambridge astronomers began to doubt their ETI hypothesis. Since the two pulsars were found to be separated from one another by over 4,000 light-years, they figured that if they were ETI transmitters, they would necessarily have been built by different civilizations. But then it seemed to them very unlikely that more than one civilization would have chosen to communicate with us at this particular point in time and in addition use a similar method of sending precisely timed pulses.

Fearing they would be inundated with reporters if their discovery became known to the public, the astronomers kept their find a tightly guarded secret until February, when they submitted a paper about it to *Nature* magazine.[2] Their paper, however, avoided making an extraterrestrial intelligence (ETI) interpretation. Rather, they proposed that these signals might be emitted from the surface of a highly dense compact star, such as a white dwarf or neutron star, that was expanding and contracting, dimming and brightening, in a very regular manner. A decision to stick to their initial ETI hypothesis would most assuredly have condemned them to attacks from skeptical colleagues and would very likely have jeopardized their chances of publishing their findings in refereed journals. Besides, their study had originally been designed to investigate natural astronomical phenomena, not to search the skies for signs of extraterrestrial intelligence. In the following months, the Cambridge scientists discovered two other extremely regular pulsating sources with comparable periods of 0.253065 and 1.187911 seconds, duly named LGM-3 and -4, respectively. Later, when these sources became known as pulsars, the four were renamed PSR 1919+21, PSR 0834+06, PSR 0950+08, and PSR 1133+16.*

Multiple-source ETI communication, however, would not be all that unusual if the signals were coming from several intercommunicating civilizations, forming a kind of galactic collective or commune. In such a case, the idea of several communicators being on line and using similar methods of transmission would seem rather plausible. Today many scientists interested in the Search for Extraterrestrial Intelligence, an endeavor known as SETI, believe that such a galactic commune could very well exist. For example, the MIT radio astronomer professor Alan Barrett was one scientist who in the early 1970s was quoted by the *New York Post* as having wondered whether pulsar signals "might be part of a vast interstellar communications network which we have stumbled upon."[3] But the idea of a communication collective had not received much discussion back in 1967, and so doubts arose.

Another reason why the Cambridge astronomers began to question their ETI hypothesis had to do with the way the radio signals were being sent. Rather than being transmitted at discrete frequencies like our own radio and television stations, pulsar transmissions covered a broad range of radio frequencies. The astronomers Robert Jastrow and M. Thompson, for example, reasoned as follows:

If an extraterrestrial society were trying to signal other solar systems,

---

*PSR signifies "pulsating source of radio" and the numbers indicate the source's sky position as seen in 1950 C.E. (an added J would signify the position in the year 2000). The first four digits give hours and minutes of right ascension measured from west to east along the celestial equator and the last two give degrees of declination measured either north (+) or south (−) from the celestial equator. The celestial equator is the outward projection of the Earth's equator onto the celestial sphere.

its interstellar transmitter would require enormous power to send signals across the trillions of miles that separate every star from its neighbors. It would be wasteful, purposeless, and unintelligent to diffuse the power of the transmitter over a broad band of frequencies. The only feasible way to transmit would be to concentrate all available power at one frequency, as we do on earth when we broadcast radio and television programs.[4]

Nevertheless, the development of particle-beam weapon technology during the 1980s brought us a step closer to realizing that broadband ETI communicators are not such a far-fetched idea after all. With this technology it is possible for us today to build a space-based device capable of projecting an intense beam of high-energy electrons that would in turn generate a highly collimated laserlike radio wave beam. This particle-beam communicator would consist of two main components: a particle accelerator and a particle-beam modulator unit (fig. 1). The particle accelerator would produce a beam of high-energy electrons traveling at very close to the speed of light. The beam modulator would apply magnetic forces transverse to this particle beam causing its electrons to deflect slightly and to convert some of their forward kinetic energy into *synchrotron radiation,* electromagnetic wave emission that characteristically spans *a broad range of frequencies.*

Synchrotron radiation was first discovered in the early 1940s when physicists at the General Electric Research Laboratory in Schenectady, New York, first powered up the Synchrotron, one of the world's first high-energy particle accelerators. During its operation, they noticed

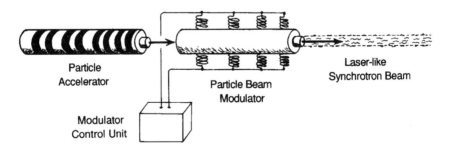

Figure 1. An ETI particle-beam communicator device I proposed that could be used to transmit pulsarlike radiation pulses to other civilizations in the Galaxy (see chapter 7 for details).

that a fascinating blue-white glow was radiating from the accelerator's high-energy electron beam. It was later found that this emission had a very broad spectrum that ranged from low-energy radio and microwaves on up to high-energy ultraviolet and X-rays. Electrons traveling near the speed of light are known to emit this broadband radiation whenever they are magnetically deflected from their normal straight-line trajectories. Their high speed causes them to emit this radiation as a narrow conical beam aimed in their direction of travel (fig. 2).

Although it was first discovered in the laboratory, synchrotron radiation was later found to be quite commonly produced in nature. Radio astronomers typically detect its presence wherever high-energy cosmic ray particles are being deflected by magnetic fields. It is detected coming from solar flare particles trapped in the Earth's Van Allen radiation belts, from cosmic ray electrons magnetically trapped in supernova remnants, and from the tremendously energetic cosmic ray barrages emitted from the luminous quasarlike cores of exploding galaxies.

The pulsed radio signals coming from pulsars have also been determined to consist of synchrotron radiation. In fact, by properly controlling its modulator unit, the particle-beam communicator shown in figure 1 could be made to produce a synchrotron beam that would flash on and off and produce a signal very similar to that coming from a pulsar. Powered by a medium-sized power plant that supplied on the order of 10 to 100 megawatts, this communicator could produce a beamed signal that even at distances of thousands of light-years would have a strength similar to that coming from a pulsar. Additional details about how such a communicator might operate are given in chapter 7.

Given that it is possible for a technically advanced civilization to produce broadband pulsarlike signals, what would be some of the advantages for them to do so, as opposed to broadcasting discrete frequency transmissions? For one thing, a broadband signal would stand

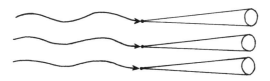

Figure 2. Electrons traveling at near-speed-of-light "relativistic" velocities emit narrow cones of synchrotron radiation when they are magnetically deflected.

a better chance of being detected by a radio telescope. Such telescopes are normally designed to receive a jumble of radio signals covering a wide range of frequencies, as would normally come from naturally occurring radio sources in the sky. A radio station broadcasting a single frequency would become lost in the background noise resulting from the thousands of radio frequencies being received. On the other hand, a broadband signal, whose intensity was made to coherently vary over all of its frequencies, would more easily stand out and be detected, and it could be detected regardless of which part of the radio-frequency spectrum an astronomer happened to be observing. If an ETI signal were instead being transmitted on a single radio-frequency channel, an astronomer would need to have the good fortune to tune in to that particular channel out of billions of available channels. It would be like trying to find a needle in a cosmic haystack. This difficulty could be overcome by retrofitting radio telescopes with specialized electronic equipment capable of rapidly processing data gathered simultaneously from millions of discrete channels. In fact, such signal-processing equipment is currently being used in the SETI program. But it is not the kind of apparatus that observational astronomers would normally use in surveying the radio-emitting sky.

Broadband signal transmissions also have the advantage of providing the recipient civilization with a way of estimating the communicator's distance. Interstellar space is filled with a tenuous medium of unattached electrons that causes lower-frequency radio waves to travel slightly slower than higher-frequency waves. This effect is due to radio-wave scattering and not to any change in the wave's velocity through space. The low-frequency radio waves from a communication pulse, then, would lag slightly behind the high-frequency waves coming from the same pulse (see fig. 3). Consequently, the recipients of the pulsed message could determine the sender's distance simply by measuring the amount of this frequency-dependent time delay. Such distance ranging would not be possible if the sending civilization were transmitting signals at only one frequency. So in retrospect, it seems that some of the reasons that were once given for discounting the possibility that pulsar signals might be of ETI origin are really not so sound.

Nevertheless, early searches for intelligent signals in space were conducted on the assumption that the transmissions would be of the discrete frequency type. The first such radio telescope search was carried out by the astronomer Frank Drake in 1959 and 1960. This project, named OZMA, used the 26-meter radio antenna at the National Radio Astronomy Observatory in Green Bank, West Virginia, to search for signals from the two closest sunlike stars, Tau Ceti and Epsilon Eridani. Figuring that ETI signals would be transmitted at a discrete frequency, they tuned their telescope to scan what they thought would be the most likely frequency, 1420.405 MHz, the 21-centimeter line wavelength that atomic hydrogen normally radiates at. However, their search turned up nothing.

Although SETI enthusiasts carried out a variety of other searches in the years that followed, there was no organized, scientifically recognized program at that time to fund such activities. Moreover, during these early years, the scientific community was not nearly as tolerant as it is today of the idea that intelligent beings might live elsewhere in the Galaxy and might even be trying to communicate with us. It was not until 1984 that astronomers first came up with irrefutable observational evidence indicating the presence of solar systems around other stars. It is not surprising, then, that back in 1967 Hewish and his Cambridge astronomy group had backed away from their ETI interpretation for pulsars.

The announcement of their findings created quite a stir in the astronomical community, for no other natural sources were then known to exhibit such precisely timed pulses. It became regarded as one of the most important astronomical discoveries of the decade. Jocelyn Bell subsequently received considerable press recognition and Drs. Hewish

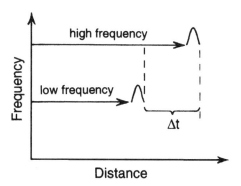

Figure 3. Compared with higher-frequency radio waves, waves of lower frequency require more time to traverse the same distance through space. By measuring the amount of delay in radio pulse arrival time, astronomers are able to estimate the distance to the pulsating source.

and Ryle, who were codirectors of the Cambridge University radio telescope project, shared the 1974 Nobel Prize in physics.

Shortly after the Cambridge pulsar findings were made public,
other astronomers began their own searches. As a result, the number
of known pulsars rose to 50 by mid-1970, to 147 by 1975, to 330 by
1981, to 550 by 1992, and to 706 by 1997, and by the end of 2005
more than 1,530 had been discovered. As many as 140 scientific papers
were published on pulsars in 1968 alone, and hundreds more came out
in the following years.

The Neutron Star Lighthouse Model

In the months after Hewish and Bell published their pulsar discovery, scientists came forth with as many as twenty different theoretical
models attempting to explain the phenomenon. The earlier suggestion
that pulsars might be radially pulsating white dwarf stars had to be
discarded following the discovery later that year of two unusual pulsars
found in the Crab and Vela supernova remnants. Both have periods
under one tenth of a second, far too short to be realistically described
by a radially pulsating dwarf star.

As an alternative, astronomers eventually settled on the neutron
star lighthouse model, which was proposed by Thomas Gold in June
1968.[5] This conceived a pulsar to be an extremely dense, rapidly rotating mass of neutrons, called a *neutron star,* which was theorized to emit
opposed beams of synchrotron radiation (see fig. 4). With each rotation, one or both of these beams would sweep past the Earth, producing
a brief radio pulse.

A neutron star is said to be formed when a star's fusion reactions
burn out, leaving the star's mass in free-fall inward gravitational collapse.
This compression is then said to be followed by a supernova explosion
whose force further compacts the star's core. The result is theorized to
be a state of matter so dense that all the star's nuclear particles have
been transformed into neutrons and packed tightly together with the
same density that exists in the nucleus of an atom. The star's core, which
initially would have a mass somewhere between 1.2 to 3 times the mass
of the Sun and a diameter about like that of the Earth, would become
compressed to a size of only 1 to 30 kilometers. If it could be brought to
the Earth's surface, one cubic centimeter of this substance would weigh
somewhere between twenty-five million and one trillion tons!

The neutron star idea was first proposed in the 1930s. But for many decades thereafter, astrophysicists were not sure whether to believe that such things really existed. It was not until pulsars were discovered that they began taking the idea more seriously, since no known natural object could explain pulsar signals. In their attempt to model pulsar signals, astrophysicists theorized that the neutron star would be spinning very rapidly, from several rpm up to hundreds of times per second, and that the resulting centrifugal forces would flatten it into the shape of a pancake.[6] It was also thought that the star would be left with a magnetic field trillions of times stronger than the Earth's magnetic field. This would somehow be frozen into the neutron star matter itself and would be typically skewed at an angle to the star's axis of rotation.

Furthermore, the neutron star was theorized to be left very hot as a result of its violent birth, its high temperature causing it to radiate a stream of high-energy electrons and other cosmic ray particles. These were thought to spew out from each of the star's magnetic poles to form two opposed, pencil-like beams. One theory suggests that the star's magnetic field would decelerate these electrons as they rushed outward and in so doing would cause them to emit two collimated beams of synchrotron radiation, the same type of radiation produced

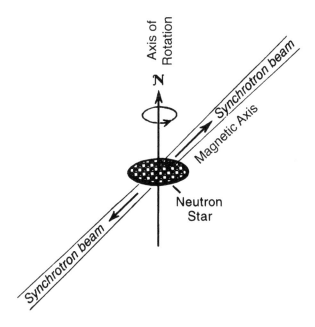

Figure 4. The lighthouse model proposed as an explanation of pulsar emission.

by our hypothetical particle-beam communicator. Since the beams were assumed to be oriented at an angle to the star's axis of rotation, they would sweep through space as the star revolved, much like searchlights from a rotating lighthouse beacon. If the Earth happened to be in the path of one or, in some cases, both of these beams, they would be observed to flash by with clocklike precision, producing a train of regularly spaced pulses.

There is, however, a fundamental problem with the neutron star lighthouse model. Whereas the model predicts that a pulsar's individual radio beeps should be evenly spaced from one another, instead *their arrival times are found to vary from one pulse to the next.* A typical example of pulse-to-pulse timing variation may be seen in figure 5, which charts a succession of pulses observed from pulsar PSR0950+08. Each horizontal trace represents the signal received over a single pulse cycle, with 260 of these cycles being stacked up for comparison. The humplike pulses consist of a rise and fall in signal amplitude that typically lasts about 9 milliseconds, or about 3½ percent of the pulsar's approximately 0.253-second pulse period. Note that each successive pulse does not occur at precisely the same phase in the pulse cycle. Instead, its timing varies with some degree of randomness. Nor is each pulse the same height as its predecessor.

Precise regularity emerges only when many pulses are averaged together to produce a *time-averaged pulse profile* such as that shown as the uppermost trace in figure 5, which has been synthesized from two thousand successive pulse cycles.* Astronomers find that the shape of this pulse contour remains amazingly constant, being virtually identical to a time-averaged profile synthesized from data obtained some days, months, or even years later. Also unlike the individual pulses, the timing of this average pulse profile is extremely precise, its leading edge always beginning to rise at its "appointed" time. When astronomers speak of the extreme precision of a pulsar's period, they are referring to the timing of the time-averaged profile, rather than to the timing of the individual pulses.

Time-averaged pulse profiles for a number of pulsars are displayed

---

*Pulsar astronomers have come to refer to the individual pulses from a pulsar as *subpulses,* and the time-averaged pulse profile they have come to call the *integrated pulse profile.* Since this terminology could be confusing to some readers, I will adhere to the terms *pulse* and *time-averaged pulse profile* when referring to these concepts.

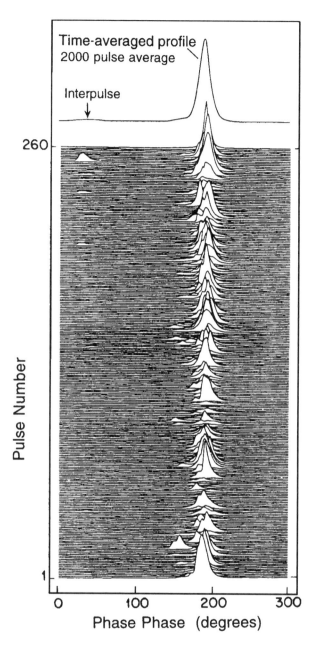

Figure 5. A sequence of 260 pulses received from PSR 0950+08. Their time-averaged pulse profile, shown at the top, is obtained by adding together 2000 individual pulses. The horizontal axis plots pulse period phase, where one complete cycle is equivalent to a cycle phase of 360° (from Hankins and Cordes, Astrophysical Journal, figure 1).

in figure 6. These are most often dominated by a single peak, but some of the wider profiles contain several peaks, or components. The pulse profiles generally span anywhere from 1 to 20 percent of a pulsar's total pulsation period, although in some cases they stretch over the entire pulse cycle.

At the time when pulsars were first discovered, radio astronomers were not aware of the pulse-to-pulse timing variability. Their radio telescope equipment was set up only to average together many pulses, typically five minutes of collected data. As a result, they were able to study only the time-averaged profile, which happened to exhibit very precise timing. Thus, it was natural for pulsar theorists to assume that the individual radio pulses being averaged together in the data output were similarly very precisely timed. The neutron star lighthouse model, with its synchrotron beam rotating with clocklike regularity, emerged in the context of this assumption *that the individual pulses were each precisely timed.*

However, within months of the time that Gold published his neutron star lighthouse model paper, investigators began discovering that there was considerable timing variation from one pulse to the next. This must be attributed mostly to the pulsar and *not* to interstellar scattering of the emitted radio waves during their journey to Earth.* This meant that the basic lighthouse model now had to be radically revised to produce *nonregularly spaced pulses.* This pulse-timing irregularity could be produced if the neutron star radiation beam were to flop back and forth as it swept through space. Alternatively, it might occur if the cosmic ray electron beam had a nonuniform substructure, its cosmic rays being concentrated into localized electron cascades, or sparks, emitted in varying directions through the beam's aperture.

But if the intervals between pulses were to vary in this manner, one would expect that when many pulses were averaged, the result would produce a profile whose shape and timing randomly changed from one pulse series average to the next. The finding to the contrary, that the

---

*In the course of their long journey, pulsar signal radio waves can become scattered away from their straight-line trajectory to Earth by intervening clouds of interstellar electrons, an effect that is most pronounced at very low radio frequencies. This can cause the pulsar signal to brighten and fade, much like the twinkling of a star. However, most pulsar observations are conducted at higher frequencies, where this scattering variability is minimized. At these higher frequencies, the majority of the pulse-to-pulse variability originates from the pulsar itself.

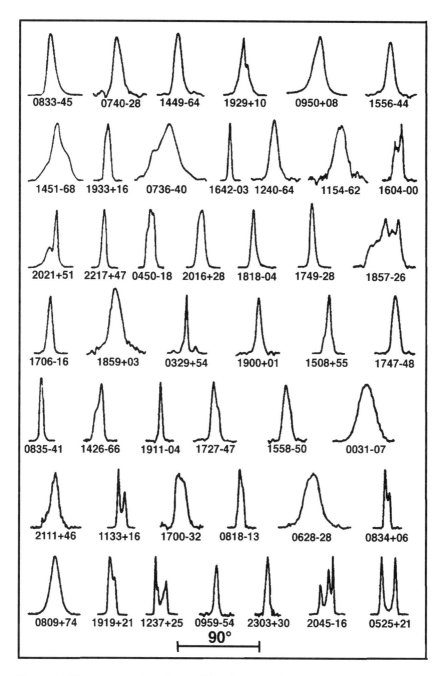

Figure 6. Time-averaged pulse profiles for 45 pulsars. The bar at the bottom indicates the length of a quarter of a pulse cycle (Manchester and Taylor, 1977, figure 1).

time-averaged pulse profile was instead very constant, indicated that the lighthouse model had to be further revised. Thus, as this lighthouse beam swept around, its cosmic ray electron sparks would have to execute their back-and-forth flickering so that, when observed over many cycles, the sequence of pulses would "paint in" a time-averaged profile that was precisely timed and had a highly ordered shape. For example, the neutron star's sparks might be imagined to pass through a kind of magnetic field "mask" that rotated together with the electron beam. This mask would have an angular width and contour similar to that of the resulting time-averaged pulse profile and would need to influence the sparks so that as they moved back and forth, the resulting sum total of their "brushstrokes" would sketch out the mask's contour. As a result, when the series of synchrotron radiation pulses were added up, they would produce an unvarying time-averaged pulse profile that matched the shape of this mask. But pulsar theorists, then, would need to explain why this postulated magnetic field mask remained so invariant all the while that the neutron star and its cosmic ray beam were spinning so rapidly and sparking in various directions. Common sense tells us that the powerful cosmic ray beam should instead "blow" the imposed mask much like a wind blowing a flame, and consequently should produce a highly erratic time-averaged pulse profile.

In summary, the lighthouse model offers a very disappointing explanation for even the most basic of pulsar signal characteristics. Its shortcomings are even further compounded when one considers the various other types of ordering that characterize pulsar signals, some of which are summarized in appendix B. One example is the phenomenon of pulse *drifting*. In certain pulsars, each succeeding pulse is observed to arrive slightly earlier than its preceding pulse, giving the appearance that the pulses are marching backward at a very regular rate and scanning across the contour of their resulting time-averaged pulse profile. To explain this kind of behavior, yet another level of complexity would need to be added to the lighthouse model. Its rotating cosmic ray beam now could no longer spark erratically in differing directions, but rather would have to produce sparks that swept across its co-rotating magnetic field mask in a highly regular fashion. Moreover, based on observations of certain other pulsars, this pulse-drifting model would have to be further revised to abruptly switch from one drift rate to another, and then to another, as if its drifting were being controlled by rudimentary rules of logic. Indeed, *the complexity of pulsar signal ordering far surpasses that of any other known astronomical phenomenon.* Even when the lighthouse model is made absurdly complex, it still falls short of satisfactorily explaining pulsar behavior, a shortcoming that today is widely recognized among pulsar theorists.

ETI Beacons?

Even after the astronomical community settled on the lighthouse model, some astronomers still entertained the possibility that at least some pulsars might be ETI beacons. Such speculation surfaced in 1974 with the discovery of PSR 1953+29, a pulsar with a very constant pulsation rate. The SETI pioneer and pulsar astronomer Frank Drake was one who suggested this pulsar might have an ETI origin. Dr. Drake is known, among other things, for having initiated Project OZMA and for having developed an equation for estimating the high probability of intelligent civilizations existing elsewhere in the Galaxy. Regarding this pulsar, he commented:

> All the other pulsars are spinning down, slowing up if you will. Not this one. We can detect no spin-down in this pulsar, making it clearly a very different beast. One other thing it can be is an intelligent civilization attempting to communicate with other worlds, because everybody has said that's how you'd mark yourself. You do something that can't be done in nature. You make the pulse rate of a nearby pulsar exactly right, not deviating in the least year after year.[7]

Observation made over the following years showed that this pulsar's period, in fact, was slowing down, but at a rate thousands of times slower than is typical of most pulsars. It was not by any means unique in this respect since astronomers have subsequently found over a dozen other pulsars that are slowing down at an even slower rate. Pulsar theorists once again came to the rescue of the lighthouse model and added special case assumptions that they believed could account for such an unusually slow loss of rotational speed. But maybe astrophysicists should not be so quick to dismiss the ETI interpretation. If extraterrestrial civilizations are attempting to communicate with us and are distinguishing their transmissions by doing "something that can't be done in nature," then pulsar signals certainly are the closest thing known to fit this criterion.

The chapters that follow present evidence that pulsars are nonrandomly placed in the sky, with particularly distinctive beacons being situated at key galactic locations that are meaningful reference points from the standpoint of interstellar communication. Of course, these seemingly intelligent sky location placements could be dismissed as just a freak occurrence, albeit a very rare one. Others may instead regard them as evidence of an underlying intelligence operating in nature. Here, we will explore the alternate theory, that pulsars are artificial beacons created by advanced ETI civilizations.

But if pulsars have been engineered by intelligent beings, the manner in which they produce their signals must somehow be explained. It is doubtful that their signals originate from space-station communicators of the kind portrayed in figure 1. Although such large-scale synchrotron beacons could be designed to produce pulsarlike transmissions, observation suggests that pulsar signals originate from near the surfaces of *star-sized bodies*.

The determination that pulsars are quite massive comes from observations made of pulsars that are orbited by a companion star or planet. In such *binary* pulsars, the gravitational pull of the circling companion causes the pulsar to describe a small orbit in space, thereby inducing a sinusoidal variation in its pulsation period. By analyzing these cyclical variations and taking into account observations of a pulsar's optically visible companion star, astronomers have become convinced that binary pulsars are indeed relatively large celestial bodies having masses comparable to that of our Sun. If we assume the hundreds of solitary pulsars that fill the sky are not much different from these binaries, we are left to conclude that all pulsar signals similarly arise from objects of rather large mass.

In such a case, does this mean that we must rule out the notion that pulsars may be ETI communication beacons? The answer is no. For example, we might imagine a scientifically advanced civilization seeking out a hot stellar core and making use of its outgoing cosmic ray electron wind for communication purposes. In other words, the electron-accelerator component shown in figure 1 would be replaced with a naturally existing stellar cosmic ray source. By using advanced technology of the sort described in chapter 8, magnetic fields might be artificially generated near the star's surface that would, in turn, decelerate the star's cosmic ray electrons and cause them to produce one or more beams of synchrotron radiation (fig. 7). By modulating these fields, synchrotron pulses could be produced similar to those coming from pulsars. As in the particle-beam communicator, these beams would be stationary and would be accurately targeted on remote locations. Depending on their degree of divergence, the beam diameters at their remote destinations might vary from the size of a typical solar system (100 times the diameter of the Earth's orbit) to about 100 light-years capable of encompassing many star systems.

These communication beacon cosmic ray sources need not necessarily be neutron stars; they could also be larger-diameter bodies such as white dwarfs or X-ray stars. It is known that such hot stellar cores, which are a common late stage in the evolution of a star, typically radiate copious quantities of cosmic ray electrons. They are much larger in diameter than neutron stars (~20,000 kilometers, rather than 1 to 20 kilometers) and are about a million times less dense, their interiors consisting of closely packed atoms rather than of closely packed neutrons. As mentioned earlier, when pulsars were first discovered, astrophysicists theorized that their emissions might be produced by white dwarfs that were either rapidly rotating or radially pulsating. But when short-period pulsars such as the Crab and Vela pulsars were discovered, the white dwarf models had to be abandoned since, due

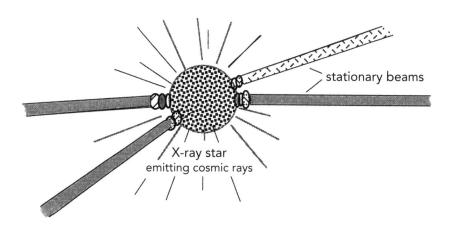

Figure 7. A stationary-beam pulsar beacon. Fields engineered near the surface of a cosmic ray–emitting star project radio synchrotron beams to targeted locations. See chapter 7 for further details.

to their relatively large size, they could neither rotate nor radially oscillate rapidly enough to explain such rapid pulsations. As a result, pulsar astronomers adopted the neutron star model as the only feasible alternative. However, if pulsar signals are of ETI origin, produced by artificially modulated magnetic fields, rather than by natural mechanical movement of the star itself, a much larger size hot stellar core could also serve as a satisfactory cosmic ray source. The previous mechanical limitations would no longer apply. More will be said about this ETI beam model in chapter 7.

Let us next examine what characteristics besides the precise timing of the time-averaged pulse profile might lead us to conclude that pulsars are communication beacons operated by intelligent beings.

TWO

A GALACTIC MESSAGE

The One-radian Marker

Scientists of a communicating galactic civilization have to deal with the problem that their message must surmount an inherent language barrier. They might solve this by framing their message in terms of universal symbols that would be understandable to any advanced civilization. Easily recognizable mathematical or geometric relationships, laws of nature, or major astronomical points of interest might serve as common reference points where a meaningful conversation could begin.

One key location in our own galaxy that would be an obvious candidate as a reference point is the Galactic center, the fulcrum mass around which the stars of the Milky Way orbit. All civilizations in our galaxy would perceive this unique site as a central location. NASA scientists referenced this Galactic center location in the space plaque message that was sent out in 1972 onboard the Pioneer 10 spacecraft. It was hoped that an alien civilization capable of space flight might one day come upon Pioneer 10 after the craft had journeyed far outside the solar system and discover the message etched on the gold-plated aluminum plaque affixed to its side (see fig. 8). The long horizontal line extending to the right of the "starburst" diagram on this plaque indicates the solar system's relative distance from the Galactic center, while the other binary coded lines radiating in radial fashion indicate the directions, relative distances, and pulsation periods of fourteen prominent pulsars. By means of these celestial markers, it is hoped that an alien civilization might triangulate the spacecraft's point of origin and thereby locate our planet. As will be seen shortly, the pulsars that this diagram innocently uses as reference points may be part of a network of ETI navigation beacons that may themselves be transmitting a message to us.

To receive the Pioneer 10 message, the recipient civilization would actually have to intercept the spacecraft and recognize that its plaque carried a message. However, suppose that a civilization were to attempt to communicate by beaming flashing radio signals through interstellar space. How might beings communicating in this way use the Galactic center reference so that their signal was not mistaken for a natural radio source? One method could be to carefully choose the direction in which they beamed their signal so that the recipient civilization would see it coming from a location that was uniquely positioned relative to the Galactic center, a location that had a particular geometrical meaning easily recognizable to the recipient.

In fact, a careful study reveals that certain very distinctive pulsars are located at positions that point out key locations relative to the Galactic center. Consider figure 9, which plots the sky locations of a

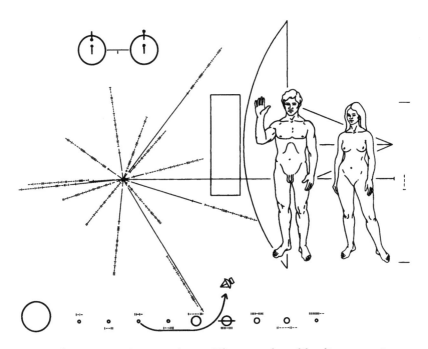

Figure 8. The Pioneer 10 space plaque. The central starlike diagram points out in relation to the Earth the locations and directions of the Galactic center and 14 pulsars.

set of 1,533 pulsars on a standard galactic coordinate map. The map's coordinate lines measure galactic longitude (ℓ) and latitude (b) much like the longitude and latitude lines on a world map.* Our Galaxy's spiral arm disk would stretch out along the map's equator with the Galactic nucleus being located at the map's center. We can get an idea of how the Milky Way's edge-on spiral arm disk and central bulge would appear on such a map by referring to figure 10, which presents a similar mapping of the distribution of the Galaxy's diffuse infrared emission.

As a whole, the pulsars plotted in figure 9 tend to congregate toward the galactic equator, the midline of the spiral arm disk. However, as we proceed to the left toward higher galactic longitudes, it is apparent that the number of pulsars drops precipitously past ℓ = 57±1°. We can get a

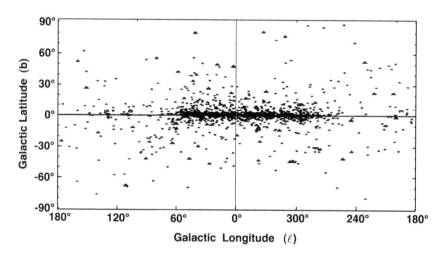

Figure 9. A galactic coordinate map showing the distribution of 1,533 pulsars (plotted from the ATNF pulsar database). The central horizontal line represents the galactic equator. The Galactic center is situated at the map's center. The anticenter (ℓ = 180°, b = 0°) lies at the map's left and right extremities.

---

*Convention places zero degrees longitude very close to the Galactic center and sweeps out a full 360° circle as our line of sight moves northward from there, tracking along the Milky Way's equator—the plane formed by the Galaxy's spiral arm disk. Convention also measures galactic latitude as angular deviation above or below this equatorial plane, positive angles being toward the galactic north pole situated in Coma Bernices north of the Virgo constellation and negative angles toward the galactic south pole in Sculptor. Astronomers did not know the precise position of the Galactic center at the time they formulated this coordinate system. Consequently, the Galactic center, which later was found to be located at ℓ = −0.0558°, b = −0.0462°, deviates slightly from the galactic coordinate origin (0,0).

clearer view of this cut-off by displaying our pulsar data in bar-graph form, as shown in figure 11.[1] The horizontal axis plots galactic longitude and lays out the Galaxy's 360-degree circumference in a series of 72 five-degree-wide longitude increments. The vertical axis plots for each longitude increment the number of pulsars lying within five degrees of the galactic equator. The 5° longitude increments have been chosen to begin from ℓ = 357.24°, rather than ℓ = 0°, in order to more clearly portray the sharp decline after ℓ = 57±1°. Within its ±5° latitude slice, the map charts a total of 1,010 pulsars, or about 66 percent of a total sample of 1,533 pulsars.*[2]

A threefold drop-off in pulsar population is evident past ℓ = 57.24°. Actually, if one-degree slices are made, one finds that the decline actually begins one degree from this point at about 56°. This feature is real and not an artifact of any observational selection effect. The Arecibo telescope, which has the advantage of being able to detect pulsars that are very faint and distant, has surveyed the sky nine degrees of longitude past this decline in population and has found few pulsars. Moreover, the Green Bank and Jodrell Bank radio telescopes, which

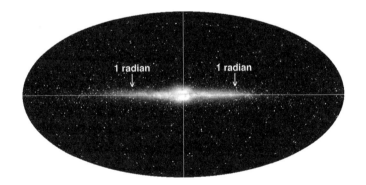

Figure 10. Sky distribution of diffuse infrared emission from the Milky Way at a wavelength of 1 to 3 microns. Data is from NASA's Diffuse Infrared Background Experiment onboard the COBE satellite (courtesy of NASA and COBE Science Working Group).

---

*Most of these detections were made in pulsar surveys that observed at a radio frequency of 400 MHz, although a portion also were detected in 1,400 to 1,500 MHz surveys. The plotted data set includes the high frequency 1992 Parkes survey, which extends from $-90° \leq \ell \leq 20°$ and $|b| \leq 4°$, and the 1992 Jodrell Bank survey, which extends from $-5° \leq \ell \leq 110°$ and $|b| \leq 1°$. These are more sensitive than the 400 MHz surveys but the latter covers only one fifth of the latitude range of our data set.

have wide sky coverage ability, have supplemented Arecibo in surveying this entire region. The sudden drop-off from a high pulsar population persists even when we plot just the brightest, and hence most easily detectable, pulsars. It also persists if we expand our latitude coordinate range to include pulsars lying within ±10° of the galactic equator.

The cut-off at this particular galactic longitude region cannot be attributed to a decline in pulsar population beyond one of the Milky Way's spiral arms because there is no arm aligned along the one-radian direction. The spiral arms instead cut across at an angle to this direction. The sudden drop-off also cannot be attributed to the molecular gas ring that circumscribes the Galactic nucleus at a distance of about 14,000 light-years. The line-of-sight optical depth of this ring reaches a maximum around $\ell = 30°$ to 40° and gradually decreases to a minimum value by $\ell = 52°$. Consequently, if pulsars were born from stars that may have formed in this ring, their populations should instead have decreased well before the one-radian cut-off.

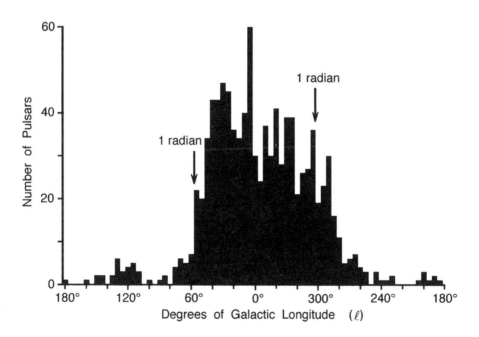

Figure 11. The number of pulsars lying within 5° of the galactic equator plotted as a function of galactic longitude. The left and right arrows indicate longitudes lying one radian from the Galactic center.

Figure 12. The 1,000-foot-diameter radio telescope at Arecibo, Puerto Rico. The Arecibo Observatory is part of the National Astronomy and Ionosphere Center, which is operated by Cornell University under a cooperative agreement with the National Science Foundation (photo courtesy of David Parker, 1997/ Science Photo Library).

The lighthouse model, which interprets pulsars as spinning neutron stars formed naturally in supernova explosions, predicts that pulsars should be distributed in the Galaxy in much the same way as supernova remnants.[3] These remnants are scattered throughout the galactic disk and increase in concentration toward the Galactic center, conforming to the distribution of the relatively massive, so-called population I–type stars, which are believed to be the progenitors of supernovae. However, the supernova remnant distribution profile shows no comparable drop-off at the one-radian longitude. The decline away from the Galactic center is more gradual, the downward slope being greatest around $\ell = 30°$ to $40°$.

From a geometrical standpoint, however, the position of this decline in pulsar population relative to the Galactic center is quite striking, for it occurs near the point on the galactic equator that lies *one radian of arc from the Galactic center;* see left arrow in figure 11. What is a radian and why is this particular galactic longitude so special from the standpoint of extraterrestrial communication? The radian is a universal concept that comes from the study of geometry. Let us begin by drawing a circle like that shown in figure 13. If we mark off a length along the circle's circumference that has the same length as the circle's radius, then the angle that subtends this arc, as measured from the center of the circle, is *one radian*. It takes a total of 2π radians to completely circumscribe a circle. Consequently, one radian will equal 360 degrees divided by 2π, or about 57.2958 degrees. Regardless of the size of the circle, this angle will always be the same.

So the radian system uses the circle's own radius as the yardstick for measuring its angles. Although not practical from the standpoint of dividing a circle into a whole number of parts, it does have the advantage that it is based on a simple geometric relation that should be known to any civilization in the Galaxy. As such, it is a prime candidate for use in extraterrestrial communication. We on Earth have chosen to use the *degree,* which divides a circle into 360 equal parts, a number

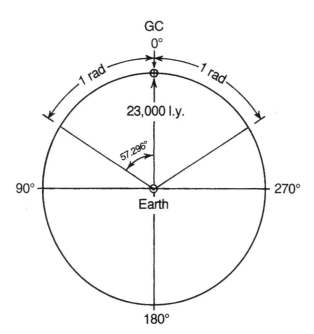

Figure 13. Illustration of the one-radian markers relative to the Galactic center (GC), as viewed from Earth.

that is easily divisible and that approximates the number of days in our solar year. But our scientists and engineers also use the radian for measuring angles, and it is the radian measure that we would choose if we were to design an ET communication message because we could be assured that it would be known to scientists of other advanced civilizations in the Galaxy who were able to receive our signals.

To see how the radian translates into measurements of galactic longitude, imagine that the circle we have drawn has a radius equal to the distance from our solar system to the center of our galaxy and whose plane is aligned with the galactic equatorial plane. The Earth would then be positioned at the circle's center, and the Galactic center (GC) would lie at a point on its circumference (see fig. 13). Galactic longitude is measured by proceeding counterclockwise around the circle from the Galactic center reference point. Since the true center of the Galaxy is positioned at galactic longitude ℓ = −0.0558°, a one-radian angular deviation from this point would lie in the northern celestial hemisphere at galactic longitude ℓ = 57.2400° (i.e., 57.2958° − 0.0558°). The northern one-radian benchmark would lie in the vicinity of the Sagitta and Vulpecula constellations and would fall very close to the sudden decline in pulsar population. A second one-radian benchmark would lie along the galactic equator at galactic longitude ℓ = 302.6484°. It would reside in the southern celestial hemisphere, near the Southern Cross constellation.

Implicit in the one-radian concept is the notion that the one-radian circumferential arc has a length equal to the radius of the circle. In the context of the galactic coordinate system, the one-radian benchmark would signify an arc length equal to our solar system's distance from the Galactic center. Since the pulsar population cut-off marks the distal end of this one-radian arc, this metaphor would symbolize a 23,000-light-year radial distance that vectors out from the Galactic center and terminates at the Earth.

One-radian Geometrical Rule

The arc length from the Galactic center to a galactic one-radian benchmark equals the distance from the Galactic center to the Earth.

By pointing out this one-radian location, the fabricators of this pulsar network would be conveying to us not only that their signals

are of intelligent origin—that they come from beings who have knowledge of circular geometry—but also that the creators of these signals
know the direction of the Galactic center *as viewed from our solar
neighborhood.*

Nevertheless, in considering such a radical proposal, many of
us would want further proof than just this abrupt decline in pulsar
population being found close to a significant geometric longitude. If an
extraterrestrial civilization had planned to mark for us this important
galactic direction, it seems that it would have done so in a more obvious
and attention-getting manner, and one that made a more precise reference to this longitude. In fact, there is a marker that exactly fits this
description. It is the most distinctive pulsar in the sky—the Millisecond
Pulsar.

The Millisecond Pulsar Marker

Of all pulsars in the sky, there is one very unusual pulsar that also lies
closest to the Galaxy's northern one-radian point, the point that marks
an angle of one radian from the center of the Galaxy. This is the pulsar
designated PSR 1937+21; see figure 14. It comes exceedingly close to
this key location, deviating by just 0.4 degree of arc! Its galactic coordinates are $\ell = 57.5089°$, $b = -0.2896°$, as compared with $\ell = 57.23995°$,
$b = 0°$ for the Galaxy's northern one-radian point. The pulsar's galactic
longitude coordinate, projected onto the galactic equator, deviates from
this one-radian longitude by just 0.27 degree, equal to about half the
diameter of the full moon.

If the ETI hypothesis is correct, we might imagine that the civilizations that sited these beacons would want to ensure that the pulsar
positioned closest to one of the Galactic one-radian longitudes would
have unique characteristics. In fact, pulsar PSR 1937+21 ranks as one
of the most attention-grabbing of all known pulsars. It happens to be
the fastest pulsar in the sky. It flashes about 642 times per second,
each of its cycles spanning just 1.557806468197941 thousandths of
a second.[4,5] If its flashes could be heard, they would sound a note of E
above high C. In addition to pulsing very rapidly, its pulsation period
is unusually constant, increasing by just 3.3 trillionths of a second
each year. With its timing characteristics presently being known to 17
significant figures, this beacon now *surpasses the best atomic clocks in
its precision.*

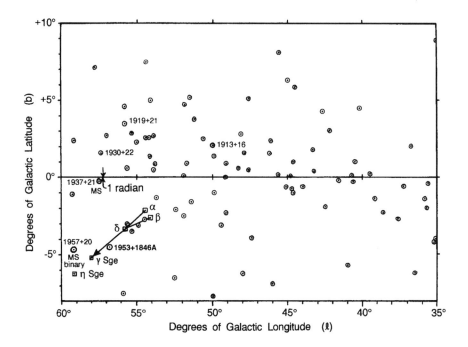

Figure 14. A close-up view showing the positions of pulsars lying between ℓ = 43° and ℓ = 57.5°. The positions of five stars forming the constellation of Sagitta (the Celestial Arrow) are marked by squares.

Most pulsars have much slower pulse periods, ranging from tenths of a second to a few seconds. However, there are a small fraction that pulse very rapidly, so rapidly that they stretch the spinning neutron star model beyond the limits of credibility. Pulsars such as these having periods shorter than 10 milliseconds (pulse rates greater than 100 times per second) have been termed "millisecond pulsars." At the time of writing, astronomers have discovered a total of about 92, and the number is slowly rising.

Because of its unique status as "chief" of the millisecond pulsars, PSR 1937+21 has been formally named the Millisecond Pulsar. Besides being the fastest pulsar, the Millisecond Pulsar is unique in several other ways. At radio frequencies it is the most luminous of all millisecond pulsars. Compared with other pulsars in the millisecond category, it emits 10 to 100 times more energy. At its distance of 11,700 light-years, it is the second brightest millisecond pulsar in the sky, surpassed by one that is brighter only because it happens to lie 25 times closer to us. Moreover, in step with its radio frequency pulses, it also emits flashes of light that are visible with optical telescopes. Optical flashing is a very rare phenomenon. Up until now it has been *observed in only four other radio pulsars,* two being the Crab and Vela pulsars, which mark the locations of the Crab and Vela supernova remnants; see chapter 5. The other two are a pulsar in the constellation of Gemini and an extragalactic pulsar that lies 160,000 light-years away in the Large Magellenic Cloud. Among millisecond pulsars, PSR 1937+21 is the only one emitting optical pulses.

The Millisecond Pulsar is also special in that it is *one of only ten pulsars known to emit giant pulses,* the Crab and Vela Nebula pulsars being also among these ten. Giant pulses are rare pulses whose intensity exceeds the average pulse intensity by manyfold. In the case of the Millisecond Pulsar, about one radio pulse in ten thousand is more than twenty times more intense than the norm and one pulse in 800,000, coming about once every twenty minutes, exceeds the norm by one hundred times. Occasionally, it produces giant pulses that rise up to 1,000 times its mean pulse intensity. When emitting a giant pulse of such high intensity, the Millisecond Pulsar becomes the brightest pulsar in the sky at radio frequencies, thereby allowing it to be easily detected by recipient civilizations. The distinguishing features of emitting both optical pulses and giant pulses single out this pulsar as an appropriate marker beacon, one that flashes its impetuous warning 642 times per second to get our attention to this key galactic location that lies one radian from the Galactic center.

Another feature that distinguishes the Millisecond Pulsar as a marker beacon is that it hardly moves from its sky location. Due to their characteristic velocities, pulsars move relative to distant background stars, as do most nearby stars. Astronomers call this *proper motion.* As of the time of writing, proper motions have been determined for 233 pulsars, and, of these, the Millisecond Pulsar is found to have the lowest proper motion. Its position changes by only 0.8±2.0 seconds of arc per millennium.[6]

In summary, then, several features together make the Millisecond Pulsar a likely candidate for a marker beacon. It is the fastest pulsing pulsar; it emits optically visible pulses that facilitate easy detection; it is the most luminous of all known millisecond pulsars; it regularly produces high intensity pulses called "giant pulses"; its time-averaged pulse profile exhibits extreme regularity; and it has a very low proper motion.

The odds that any one given pulsar will come within 0.4 degree of the one-radian benchmark are about 1 in 14,300.* However, the odds that this pulsar would also be one that emits optical pulses is 5/1533, or about one chance in 300; and the odds that this pulsar would also be one that emits giant pulses would be 10/1533, or about one in 153. As a very conservative estimate, multiplying these probabilities we find that the chance that a pulsar as unique as the Millisecond Pulsar was the one to hold this honor would be almost one chance in a billion! The odds are overwhelmingly against nature having arranged this placement. If we had to choose between coincidental placement near this key galactic location and purposeful placement by a galactic civilization, the odds seem to lean heavily in favor of the ETI alternative. That is, an ET civilization may have purposely sited this beacon here to mark a one-radian-arc deviation from the Galactic center that would be evident from our particular viewing perspective. In this way, by employing the universal language of geometry, they would be alerting our astronomers to the artificial nature of their signals.

But seeing that this pulsar has such a low velocity in the plane of the sky, if it was intended to mark for us the location of the galactic one-radian point, why does it deviate by 0.4 degree from this location? Was it not possible for its creators to find a stellar cosmic ray source that was better positioned? Later we will discover that this angular deviation was intended and that, in fact, we are being notified of its amount.

It is important to note that the Galaxy's northern one-radian point is situated near the ancient constellation of Sagitta, which portrays the Celestial Arrow. Sagitta's arrow tip is represented by the star Gamma Sagittae (γ Sge), which is located in the sky at $\ell = 57.9661°$, $b = -5.2066°$. Of all visible constellation stars, the sky position of this one comes nearest to this one-radian benchmark, deviating in longitude by just 0.73 degree. In fact, there is strong evidence to suggest that the ancient astronomers who outlined the Sagitta constellation knew the location of the Galactic center and purposely marked this one-radian point. Consequently, in terrestrial star lore, Gamma Sagittae stands as the counterpart for the Millisecond Pulsar, both of which designate this key galactic location.

The Vulpecula pulsar (PSR 1930+22) is also of interest (see fig. 15). Being positioned in the Vulpecula constellation at $\ell = 57.3435°$,

---

*Here we take the ratio of the area of a circle of radius 0.4 degree to an area of the sky measuring 360° by 20°, assuming that most of the pulsars lie within ±10° of the galactic plane. This calculates as 0.5/7200 = 7 × 10^{-5}.

b = +1.5467°, it is the *second* closest pulsar to the northern galactic one-radian point. It is also the *closest* pulsar to the longitude meridian that passes through this equatorial one-radian point, deviating from this meridian by just 0.1 degree. Interestingly, the Millisecond Pulsar and the Vulpecula pulsar, which are among the three whose longitudes come nearest to the northern galactic one-radian point longitude, are so positioned that a trajectory line drawn through their locations currently passes within just 0.01 degree of arc of Gamma Sagittae, the pointer star of the Celestial Arrow.* It is difficult to interpret this close alignment as

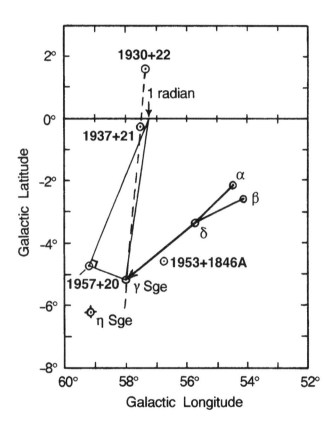

Figure 15. A sky diagram showing the location of the Millisecond Pulsar (PSR 1937+21), Eclipsing Binary Millisecond Pulsar (PSR 1957+20), the Vulpecula pulsar (PSR 1930+22), and the Sagitta pulsar J1953+1846A relative to the stars that form the constellation of Sagitta, the Celestial Arrow.

---

*The alignment would have been exact around one millennium ago, but since then has worsened due to Gamma Sagittae's gradual lateral movement, or "proper motion."

just a coincidence. If this pulsar indicator is an artificial beacon intentionally placed by a galactic collective of intercommunicating civilizations, then they would have had to know how the night sky appears from our particular galactic locale—that of the bright stars nearest our solar system, Gamma Sagittae is the only one seen from our general locale to come closest to the galactic one-radian point, a benchmark whose sky position is determined by our particular viewing direction.

As is more fully explained in the book *Earth Under Fire,* the Sagitta constellation is part of an ancient constellation lore message that uses the language of archetypal metaphor to record the past occurrence of a Galactic core explosion.[7] This message relates how an intense cosmic ray wind issued from the Galactic center and after a 23,000-light-year journey began to pass through our solar system close to the end of the ice age. Sagitta symbolizes this intense volley of cosmic ray particles that journeyed from the Galactic center to the solar system, a distance symbolically equivalent to Sagitta's one-radian-arc flight away from the Galactic center along the galactic equator. More will be said about this in chapter 4.

The Eclipsing Binary Millisecond Pulsar

PSR 1957+20 is another distinctive millisecond pulsar that happens to be located in this key part of the sky and which in many respects is in itself unique.[8] It is situated in the Sagitta constellation and lies 5,000 light-years from us. It is positioned at $\ell = 59.1970°$, b = −4.6975, just beyond the longitude where the pulsar population drops off sharply. Like the Millisecond Pulsar, it excels as a galactic time standard. It ranks twenty-first among all pulsars in constancy, its period increasing by just one half-trillionth of a second each year. However, what makes it most unusual is that it is the *second fastest* pulsator of the pulsar family. Astoundingly, its period of 1.60740168480632 milliseconds exceeds the length of the Millisecond Pulsar's period by just 3.1837 percent.

It is indeed very unusual that the two most rapid pulsators in the Galaxy would not only have periods *so nearly identical,* but that they also would be positioned in the sky *so close together,* within 4.5 degrees of one another, *and so near this galactic one-radian point.* Astronomers who have studied these two millisecond pulsars have made no comment about the proximity of these pulsars to this key geometrical point, probably because they have not seriously considered the ETI connection. Nevertheless, the close mutual proximity of these two unusual sources has caught the attention of some. It inspired the astronomer Dan Stinebring to jokingly refer to this region as "Pulsar Heaven" because, as he jestingly put it, "that seems to be where the most interesting stars go after they die."[9] In alluding to pulsars as "dead stars," he presumes, as do the majority of astronomers, that a pulsar is a spinning neutron star, hence a stellar core remaining after the explosive death of its progenitor star. This standard view of neutron stars is a bit off the mark. As we will see later on, neutron stars are not dead remnant cores, nor are they necessarily formed in supernova explosions.

PSR 1957+20 differs from the Millisecond Pulsar in that it is a *binary* pulsar. Its 1.4 solar mass neutron star is orbited by a companion star, in this case a 0.02 solar mass white dwarf, and this introduces a sinusoidal modulation in the timing of the pulsar's pulses. The cosmic ray wind from the central neutron star continually blows gas off the surface of its orbiting dwarf companion, causing the two to be enveloped in a small nebula (see fig. 16). Interestingly, studies of the pulsar's proper motion show that it is moving toward the Galactic center. This relative motion through the interstellar medium causes a wind to blow toward the nebula from the Galactic center direction. As a result, the nebula's upwind, Galactic-center side is compressed into a bow shape and its leeward side is blown outward in a cometary fashion.

Astronomers are able to determine the orbital characteristics of a binary pulsar to a high degree of accuracy by observing the cyclical increase and decrease in its period that occurs as the pulsar and its companion star orbit one another. PSR 1957+20, for example, has been found to describe a highly circular orbit having a radius of 26,770 kilometers. Its orbit is so precisely formed that it *deviates from perfect circularity by less than one part per billion!* By comparison, the Earth's orbit around the Sun deviates from perfect circularity by more than one part in 10,000.

Another noteworthy aspect of this binary millisecond pulsar is that its companion dwarf star periodically passes in front of it, temporarily occluding its signal. Of the 112 binary pulsars, *only 14* have their orbital planes directed toward our solar system so that their partner star periodically blocks the pulsar's signals. PSR 1957+20 happens to be one of these. Every 9.2 hours its partner occludes its signals for about 50 minutes. Moreover, of the known eclipsing binary pulsars, this one has the most circular orbit. Because it is unique like the Millisecond

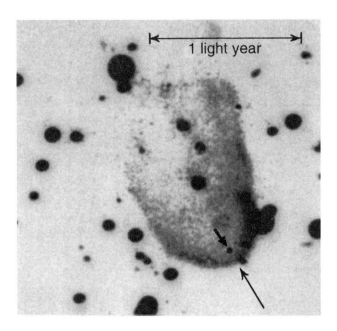

Figure 16. The cometary nebula around the binary millisecond pulsar PSR 1957+20 and its companion star. The position of the companion is indicated by the small downward arrow. The superimposed upward arrow indicates the direction of approach for an interstellar wind blowing from the Galactic center direction (Kulkarni and Hester, Nature, figure 3; photo courtesy of S. R. Kulkarni).

Pulsar, we will distinguish it with the proper name *Eclipsing Binary Millisecond Pulsar,* or EBM Pulsar for short.

The eclipsing phenomenon indicates that we lie quite close to the EBM Pulsar's orbital plane—or, in other words, that the binary's orbital plane is oriented edge-on relative to us. Is it just chance that this unique pulsar, the second fastest pulsator in the sky, *happens to trace out an orbit aimed in our particular viewing direction,* or could this be prearranged through intelligent design, an indication that this message is meant for us? To further call our attention to its location, this pulsar, like its distant partner the Millisecond Pulsar, emits giant pulses. So it too is one of just 10 out of 1,533 pulsars known to emit giant pulses.

Noting that the unique Millisecond and EBM Pulsars have not only proximal sky positions but also pulsation periods that are unusually close to one another, we are left to wonder whether the two are somehow symbolically related. If the two pulsars have pulse periods so close together, was something intended by their 3.1837 percent deviation? Let us for the moment suppose that they are meant to tell us something and see what we can discover. Instead of taking the difference between the pulsars' periods, let us take the difference between the galactic longitude coordinates of the two pulsars: in other words, the positions that would be marked out along the galactic equator if lines were drawn from each pulsar perpendicular to the galactic plane. This difference calculates to be 1.68807 degrees (i.e., 59.19697° − 57.50890°). Seeing that one of these pulsars, the Millisecond Pulsar, closely marks the Galaxy's northern equatorial one-radian point, we are enticed to express this small angle in radians. As mentioned at the beginning of this chapter, the radian is the angular measure of choice for communicating with other galactic civilizations since, being based on geometry, it would be universally understood. So, dividing 1.68807° by 57.2958° (one radian), we get 0.02946 radian, or 2.946 percent of a radian.

Interestingly, this comes very close to 3.18 percent, but it is not quite large enough. The thought occurs that if only the Millisecond Pulsar had been placed a little bit closer to the one-radian point, we might have gotten the correct answer. But now we notice the 0.26895-degree deviation between the Millisecond Pulsar's galactic longitude and the one-radian point longitude (i.e., 57.50890° − 57.23995°). Recall that earlier we wondered why this pulsar was not more accurately placed closer to the galaxy's one-radian point. However, suppose this discrepancy was intentional. Bisecting this 0.26895-degree arc segment ($\overline{AB}$ in fig. 17) yields 0.13448 degree and adding this to the 1.68807-degree difference in pulsar longitudes ($\overline{BC}$ in fig. 17) yields 1.82255 degrees. Now if we express this in radians by dividing the sum by a one-radian arc of 57.2958 degrees ($\overline{AO}$ in fig. 17), we find that this arc length is 3.1809 percent of a radian, which comes very close to the 3.1837 percent difference in the periods of these two millisecond pulsars. The two percentages differ by just 0.09 percent. But, as already noted, the EBM Pulsar has a proper motion toward the Galactic center, so in earlier years this longitude ratio would be calculated to be closer to this 3.1837 percent value. Knowing that the pulsar's galactic longitude is decreasing at the rate of 26±6 milliarc seconds per year,[10] if we use the longitude position that the EBM Pulsar would have had around 1800 c.e., which was 5.4 arc seconds (0.0015°) farther from the Galactic center, this ratio would calculate to be exactly 3.1837 percent.

On the assumption that these two pulsars have an ETI origin, their signals would necessarily have been intended for our solar system, since from another vantage point—say, from the nearest star system four light-years away—the resulting parallax would produce a sizable deviation in the pulsars' relative positions. This would have deteriorated the close correspondence presently seen between their angular separation, which when measured in radians indicates the fractional difference in their pulsation periods. Again we may choose the alternative hypothesis that it is just a matter of chance that the Millisecond Pulsar happens to have this position relative to the galactic one-radian point on the one hand and the EBM Pulsar on the other, but the probability of such an arrangement is quite small. The probability figures to be one chance in a billion that this one unique millisecond pulsar, the most rapidly pulsing pulsar in the sky, should come within 5.4 arc seconds of the point where this particular angular ratio is displayed (the area of a circle of radius 5.4 arc seconds divided by an area measuring 360° by 20°). As before, considering that this pulsar is one of only five that emit optical pulses and one of only ten that emit giant pulses, the chances of finding this "marker" pulsar here calculates to be about one in 50 trillion!

But there is more. As before, we are enticed to look at the relation of these two pulsars, the two fastest pulsing pulsars in the sky, both of which also happen to be irritatingly close to one another. Since one of these pulsars stands as a marker of the Galaxy's one-radian point, let us compare the angular separation of the galactic longitudes of these two pulsars, $\overline{BC}$ = 1.68807°, to the angular separation between the galactic longitudes of the Millisecond Pulsar and the galactic one-radian point, $\overline{AB}$ = 0.26895° (see fig. 17). We find that they are in the ratio 1 to 6.2765. This comes extremely close to the ratio 1 to 2π where 2π equals approximately 6.2832. In fact, it deviates by only 0.10 percent. If the EBM Pulsar and the Millisecond Pulsar had been separated by just 0.0018° more (an additional 6.5 arc seconds), this ratio would have been exactly 2π. This is almost the same discrepancy we noted in the indication of the ratio of the pulsar periods. So taking into account the proper motion of the EBM Pulsar, we find that somewhere around 1750 c.e. the ratio of these two segments would have been exactly 2π. Or, put another way, if the segment $\overline{BC}$, the longitude separation of the two millisecond pulsars, is interpreted as the 360-degree circumference of a circle, then the arc $\overline{AB}$, which indicates the separation of the Millisecond Pulsar from the galactic one-radian point, would on that date measure exactly 0.15915 of $\overline{BC}$, or one radian. How appropriate. And, with amazing relevance to the concept being portrayed, the Eclipsing Binary Millisecond Pulsar describes an almost perfect circle in

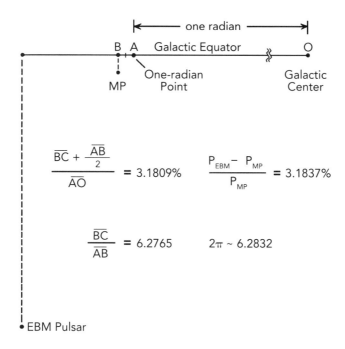

Figure 17. Positions of Millisecond Pulsar and EBM Pulsar in the vicinity of the galactic one-radian point showing how the layout of their longitudes depicts the fractional difference between these pulsar periods (P_i) as well as the pi ratio.

the sky as it orbits its companion, the most circular orbit of all eclipsing binary pulsars.

So, this millisecond pulsar "constellation" is apparently telling us the one-radian concept in a variety of ways, its one-radian ratios being depicted most accurately between about 1750 and 1800 C.E.* If we want to calculate the probability that this portrayal is a result just of a chance placement of these pulsars, we must take into account the probability that the EBM Pulsar would be so carefully placed in relation to the Millisecond Pulsar. Figuring the probability that the second fastest

---

*At the time I had published the first edition of this book in 2000, I had not yet realized that the relative sky positions of these two unusual millisecond pulsars were portraying these unique ratios. But I had been left with the pestering feeling that there was more to be discovered in regard to the Millisecond Pulsar's very small deviation from the Galaxy's northern one-radian point and in regard to the small deviation between the periods of these two pulsars. These discoveries had to wait five years, and came while I was in the process of editing this second edition for publication.

pulsar would be uniquely positioned with an accuracy of the order of 6.5 arc seconds, which calculates to one chance in 700 million, and considering in addition the rarity that it would be an eclipsing binary pulsar and, in particular, of having the most circular orbit of all eclipsing binary pulsars, we estimate a probability of one chance in a trillion. Taking into account that the EBM Pulsar has the rare characteristic of producing giant pulses, we calculate an overall probability of one chance in 185 trillion.

But we must also take account of the probability that its partner pulsar, the Millisecond Pulsar, would be sited where it is to portray this ingenious message. So, we must multiply this probability by the probability of one in 50 trillion that we calculated earlier. This gives the exceedingly small number of *one chance in 10^{28} that this arrangement of pulsars is due to chance*. Apparently, the odds weigh increasingly in favor of the ETI interpretation.

The EBM Pulsar is positioned at a key location that calls attention to the one-radian concept in yet another way. Namely, of all pulsars, its sky position comes closest to that of Gamma Sagittae, the constellation star that marks the tip of the Celestial Arrow. In fact, this pulsar is so positioned at $\ell = 59.1970°$, $b = -4.6975°$ that its sighting through the pointer star Gamma Sagittae makes almost an exact right angle with respect to its sighting through the galactic one-radian point (see fig. 15). Presently, the angle between these trajectories is 89.85 degrees, or essentially a right angle. In addition, the pulsar's galactic longitude is almost the same as that of Eta Sagittae, the "target star" toward which the Celestial Arrow is flying, deviating by less than a minute of arc.

The seemingly careful positioning of the two fastest millisecond pulsars and the Vulpecula pulsar (PSR 1930+22) relative to the Gamma Sagittae arrowhead star is quite astounding. The possibility presents itself that the network, or federation, of extraterrestrial civilizations that engineered these pulsars had carefully chosen the positions of these pulsars knowing in advance the layout of stars in our part of the Galaxy, stars bright enough to be visible to us in the nightime sky. Knowing this, they may have decided to mark the one star that would be visible to the human naked eye whose galactic longitude coordinate came closest to the galactic equator one-radian point. Or, this intriguing association could be explained if members of this galactic ETI network visited our planet long ago, had contact with our ancestors, traced out for them the Sagitta constellation, as well as other asterisms, and instructed them to pass down this extraterrestrial star lore through the generations.

But considering all of this together, we are left with a major philosophical problem. These unique pulsars exhibit these sophisticated geometrical relations—that is, the pi ratio and the one-radian concept—only if they are viewed from our particular solar system. From the vantage point of any other star system, parallax effects would cause these stars to have very different sky positions relative to one another and relative to the position of the galactic one-radian point seen from that location. Consequently, we are inevitably led to conclude that *this message is meant for us, for residents of our solar system.* But, we wonder, why would a civilization go to so much trouble to engineer entire star systems just to communicate with us? It challenges the imagination just to think that a civilization would take the trouble to put in place the necessary kind of equipment that would project force fields to the surface of neutron stars in such a manner as to create these flashing beacons. But this is a small matter in comparison with other considerations, such as the siting of these sources. For example, it is unlikely that the required cosmic ray–emitting neutron stars just happened to be residing at these two appropriately placed locations. We must conclude that they would have had to have been transported there, somehow moved from their initial homes and purposely sited at these spots. If they have used such stellar power sources, they would be dealing with masses at least 1.4 times the mass of our Sun. How did these galactic artisans move such large masses? Moreover, how did they manage to orient the orbital plane of the Eclipsing Binary Millisecond Pulsar and its orbiting white dwarf companion to ensure that the plane would be aimed in our direction? Furthermore, how did they influence these bodies to ensure that their orbits about one another would be so perfectly circular? The companion star orbiting this pulsar has a mass about 2 percent of the Sun's mass or about 23 times the mass of Jupiter. What beings could have performed such an immense engineering feat as this? The powers required to accomplish this go beyond anything we can conceive. They approach the supernatural.

At this point we are left with the insistent question: Could this all be orchestrated by a Divine Being? Could the Universe be intelligent and be going out of its way to let us know that its collective Intelligence exists, and, more important, that It cares about us? For, as we will see further on, the message being conveyed is attempting to warn us about an impending galactic cosmic ray disaster, one that could have a major effect on our planet. Alternatively, could the creators of these pulsars be living beings, albeit beings from a civilization of unimaginable spiritual advancement, who command powers similar to those attributed to the gods of old? The saying comes to mind that with faith you can move a mountain. Perhaps there exist beings who through their faith can will the movement of entire star systems.

Unlike a UFO encounter, which is transitory and leaves behind little evidence that could be used later to convince skeptics, pulsars are continuously in the heavens sending their signals. Their data is well documented in scientific journals. But when properly and objectively studied, this data inevitably leads to the conclusion that a galactic civilization of unusually high advancement does exist and is attempting to communicate with us. With so many Hollywood movies portraying aliens as being out to destroy us, it is a relief to find that the truth may actually be quite different. In his book *Exopolitics*, Alfred Webre convincingly argues that a federation of galactic civilizations does exist, that they conduct themselves in accordance with a galactic code of law, and are led by a civilization or group of civilizations that are benevolent and very spiritually advanced.[11] It stands to reason that the powers of good would ultimately dominate in the evolution of a galactic society, for technology always proceeds faster in societies that have made peace with their neighbors and whose populace regularly derives inspiration from the higher levels within.

Another puzzling aspect to consider is the speed-of-light problem. The Millisecond Pulsar is estimated to reside at a distance of 3.6 kiloparsecs (11,700 light-years) while its sister pulsar, the EBM Pulsar, resides at half this distance, about 1.53 kiloparsecs away (5,000 light-years). But since their pulsed radio signals travel through space at the speed of light, the radio pulses we are seeing today originated from the Millisecond Pulsar 11,700 years ago and from the EBM Pulsar 5,000 years ago. This implies not only that this communication effort has spanned at least 12,000 years, but that it has involved an immense amount of foresight and planning as well, for even though the two pulsars are separated by almost 7,000 light-years, their creators have succeeded in ensuring that their beamed signals would together convey a coherent message illustrating the one-radian concept.

There is also the puzzling aspect that the pulsars' message is transitory. Their geometric relations would have been most accurately portrayed if viewed between 1750 C.E. and 1800 C.E. and since that time the portrayal accuracy has been gradually declining, as the EBM Pulsar is in motion and has been changing its longitude coordinate at the rate of 2.6±0.6 arc seconds per century. So, if for some reason our technological development proceeded slower than expected and we had instead discovered pulsars for the first time five thousand years from now, the apparent sky positions of these pulsars would no longer be accurately portraying these relations. Around the beginning of the nineteenth century, classical physicists such as Coulomb, Oersted, Ampère, and Faraday were laying the foundations for electromagnetic theory. Over a century later, in 1937, Grote Reber launched the era of radio astronomy with the construction of the first radio telescope. Pulsars were discovered about thirty years after that. Is it fortuitous that these scientific and technological developments occurred at this particular time in our history so that we would have the benefit of appreciating the pulsar message so carefully and painstakingly spelled out for us in the sky? Did the pulsar builders 12,000 years ago somehow foresee that Earthlings would have developed radio telescope technology around the targeted time of the arrival of their message, plus or minus several hundred years? Or, could it be that human technological evolution is not left just to chance, that our development is externally influenced to follow some sort of prearranged timetable? We encounter the same puzzle in interpreting the messages of the Crab and Vela pulsars, which will be discussed later.

If radio pulsars are artificially constructed beacons, the sheer magnitude of such an interstellar communication project causes one to pause. Is it possible that galactic civilizations would set up so many beacons and maintain them over such a long period just to communicate with our particular galactic region? Perhaps there might be some auxiliary purpose for their deployment, such as their use as reference points for navigating spacecraft throughout the Galaxy (see chapter 3).

Other Eclipsing Binary Pulsars

As we have seen, the Millisecond Pulsar and the EBM Pulsar display the one to 2π ratio, with the 2π measure being defined by the angular deviation of the EBM Pulsar, whose binary orbit describes an almost perfect circle in the sky. In addition, they call attention to the galactic one-radian point by acknowledging through their period difference that the deviation of the fastest millisecond pulsar from this galactic benchmark was precisely gauged. Knowing that the Millisecond Pulsar symbolizes a kind of reference point in this message, it is reasonable to divide all other pulsar periods by the Millisecond Pulsar period to see if any of these normalized periods are whole-number multiples of pi. Doing so, we find that a number of periods show up as near multiples—in fact, more than can be accounted for by random chance. For example, among the set of 1,533 normalized pulsar periods, 10 were found to be within about ±0.1 percent of being an integer multiple of pi. But, based on random chance, one would expect only 4 pulsars to come this close to a pi-multiple value.

Of these integer multiples, one stands out from all the rest—the normalized period of the millisecond pulsar J1953+1846A found in the globular star cluster M-71A. Whereas the other nine pulsars exhibit pi-multiple period ratios ranging from 57 times π to 587 times π, the period ratio of this pulsar is almost exactly pi. That is, its pulsation period (4.8883 ms) is almost exactly pi times the period of the Millisecond Pulsar (1.5578 ms), the ratio of these pulsars (3.1380) deviating by 0.11 percent from being exactly equal to pi. But if its pulse period was purposely chosen to indicate an exact pi relation, it is not yet clear why this small discrepancy should be present. Like the Millisecond Pulsar, its pulsation period is found to be very constant, so we cannot attribute the discrepancy to a gradual period change.

But there are other things about this millisecond pulsar that make it unique. It falls in the constellation of Sagitta and is the fourth closest pulsar to the galactic one-radian point, its longitude deviating by just 0.5 degree. In addition, it is the second closest pulsar to the tip of the Celestial Arrow. It has almost the same angular separation from Gamma Sagittae as the EBM Pulsar, 1.381 degrees as compared with 1.332 degrees (see fig. 15). The two pulsars would have had the same angular separations from this pointer star some millennia earlier due to Gamma Sagittae's proper motion. Furthermore, like the EBM Pulsar, J1953+1846A is an eclipsing binary millisecond pulsar with its orbital plane aimed in our direction.[12] And, like the EBM Pulsar, its orbit is very circular. It has an eccentricity of less than 0.001, meaning that its orbit deviates from perfect circularity by less than one part in a million.[13]

Moreover, it is unusual that of the 14 known eclipsing binary pulsars, two would be found with sky positions so close together, the two pulsars being separated by only 2.46 degrees. It is even more unusual that

TABLE 1. DISTINCTIVE FEATURES OF THE ONE-RADIAN POINT PULSAR MARKERS

The Millisecond Pulsar

1. Is the fastest pulsator in the sky.
2. Is the most luminous of all millisecond pulsars.
3. Is one of just five pulsars that emit optical pulses.
4. Is one of just ten pulsars that emit giant pulses.
5. Is the closest pulsar to the Galaxy's northern one-radian point.
6. Is the most stationary pulsar; has the lowest proper motion of any pulsar.
7. A line extending through PSR 1930+22, the closest pulsar to the one-radian meridian, grazes Gamma Sagittae, the constellation star that comes closest to the northern one-radian point.

The Eclipsing Binary Millisecond (EBM) Pulsar

1. Is the second fastest pulsator in the sky.
2. Its period deviates from that of the Millisecond Pulsar by just 3.18 percent. This deviation equals the difference in the longitude of this pulsar from the longitude of the midpoint between the Millisecond Pulsar and the one-radian point expressed in radians.
3. The difference in longitude between the Millisecond Pulsar and the one-radian point, when compared to the difference in longitude between the EBM Pulsar and the Millisecond Pulsar, is found to be in the ratio of 1 to 2π.
4. Lies within 4.5 degrees of arc of the Millisecond Pulsar.
5. Is one of just 14 known eclipsing binary pulsars, pulsars whose orbital planes happen to be oriented edge-on in our direction.
6. Has the most circular orbit of all eclipsing binaries.
7. Is one of just ten pulsars that emit giant pulses.
8. Is the closest pulsar to Gamma Sagittae, the one-radian point indicator star in our constellation lore.
9. A line drawn from this pulsar through the Galaxy's northern one-radian point makes a right angle with a line drawn from this pulsar through the Gamma Sagittae indicator star.
10. Is surrounded by a nebula being pushed back by a wind coming from the Galactic center direction.

the two pulsars in the sky that are closest to the Gamma Sagittae pointer star would both be eclipsing binary pulsars with highly circular orbits, and that both would be separated from this star by almost the same angular amount. It is then not surprising to find that both of these eclipsing binary pulsars symbolically encode the one-radian concept of pi.*

As discussed above, the two galactic one-radian point markers, the Millisecond and EBM Pulsars, have several distinctive features that set them apart from other pulsars (see table 1 for a summary). Together they convey a coherent message about the Galaxy's northern one-radian point whose designated position is evident only from our viewing direction. The chance is exceedingly small that through natural random occurrence two uniquely placed pulsars would develop so many attention-getting features and be so positioned to express such relevant geometric relations.

---

*Two other eclipsing binary pulsars found to have relatively close sky positions are B1744–24A and J1807–2459A, being separated by 4.38° of arc. The first, $\ell = 3.84$, b = 1.70, is located in the Terzan A globular star cluster in the vicinity of the Galactic center. The second, which is much closer to us, lies at $\ell = 5.84$, b = –2.20. Also, the Terzan A cluster contains the third and fourth fastest millisecond pulsars, both of which have periods relatively close to one another, deviating by 3.1% of their pulsation period. At present, no particular significance is attributable to these particular pulsars. It is worth noting only that they lie a few degrees from the part of the sky where our ecliptic plane crosses the galactic equator. Their alignment, however, does not specifically designate this location.

THREE

THE GALACTIC NETWORK

Superluminal Space Travel

If pulsars are ETI beacons that have been fabricated by technically advanced galactic civilizations, their existence would have some very profound implications. Not only would it indicate that intelligent life exists elsewhere in the Galaxy, but also that it exists in numerous parts of the galaxy. To carefully craft a distribution of pulsars that stretches out over a distance of more than 100,000 light-years, civilizations in these diverse locations would have to be in close cooperation over a comparable duration of light-travel time. Not only does this require a functioning Galactic communication network, a kind of "galactic Internet," but it also requires a very long-term collective commitment by these civilizations.

The task of orchestrating an organized pulsar array covering such vast distances would be possible only if these civilizations had either the means for faster-than-light space travel or the means of communicating among themselves at speeds far greater than the speed of light. If superluminal travel is possible, it is unlikely that it would be accomplished with conventional rocket technology. Rather, it would involve a method of controlling gravity fields.

One area that shows great promise in the development of such field propulsion technology is the discipline called *electrogravitics*. This field, which was pioneered in the early twentieth century by the physicist and inventor T. Townsend Brown, exploits the subtle interrelation between electric and gravitational fields. Brown demonstrated that when a capacitor is charged to a sufficiently high voltage, it will experience a gravitational thrust in the direction of its positive pole.[1-3] The aerospace industry began an intensive study of this phenomenon back in the mid-1950s, and there is an indication that the B2 bomber incorporates an electrogravitic drive based on ideas described in Brown's patents.[4,5]

Academic physics has been reluctant to acknowledge the existence of the effect since it blatantly violates the tenets of general relativity. That is, according to general relativity, only masses can exert gravitational force on other masses and they do this by warping the surrounding space-time dimension metric. Moreover, this force is always attractive. Cosmologists have now tried to resurrect the idea of the universal existence of a repulsive gravitational field, albeit an exceedingly weak one. But this is an ad hoc addition and not a prediction of classical general relativity. Even if one were to admit its existence, it is so weak that it would have no practical value for propelling a craft in space. The idea that a craft might be engineered to artificially alter its own gravity field and thereby propel itself forward independent of the effect of a local gravity field is utterly foreign to general relativity. Moreover, general relativity fails to predict any linking between gravitational and electrostatic fields. Before his death, Einstein, obviously aware of ongoing electrogravitic research, was toiling to devise a modification of his theory that would allow a unification of the two fields, but he never succeeded. Following his death some unified field theories were proposed that predicted coupling between charge and mass, but these effects were expected to be observed only at exceedingly high energies available only in the beams of atom smashers that had yet to be developed.

There is one theory, however, that does predict electrogravitic coupling at comparatively modest voltage potentials. This is the physics methodology of *subquantum kinetics*.[6,7] Subquantum kinetics predicts the existence of relativistic effects such as gravitational time dilation, gravitational redshifting, the gravitational bending of starlight, and the increase of inertial mass and slowing of clocks as velocity is increased. Thus, subquantum kinetics subsumes phenomena that have traditionally been considered under the purview of general and special relativity. But its approach is very different. For example, in subquantum kinetics, gravitational force does not result from the geometric "warping" of space-time by masses. Rather, it proposes that it is the length of an object in space that changes, not the space dimension itself, and it is the rate of a clock that changes, not the time dimension itself. Masses generate classical gravitational potential energy fields around themselves without affecting the geometry of space and the gradients in these gravity fields create forces in surrounding bodies through their effect on processes postulated to take place at the subquantum level—the etheric level. So, when engineers use subquantum kinetics to design future craft capable of traveling at superluminal velocities, they will be referring not to "warp drive," but rather to *gradient drive*.

The force field beam projector developed by the Russian scientist Evgeny Podkletnov is one electrogravitic technology that looks very promising and which could lead to the reality of superluminal space travel with a minimal amount of engineering. This device generates a columnated four-inch-diameter gravity impulse by discharging a two-million-electron pulse through a superconducting disk.[8-11] This effect is predicted by subquantum kinetics: namely, the electrons in the discharge carry with them a negative electric potential field and positive (repulsive) gravity potential field. Upon reaching the anode, the electrons come to a halt, but their associated gravity field continues to propagate forward. Podkletnov and his collaborators have demonstrated the ability of these gravity pulses to momentarily exert 200,000 Gs of repulsive force during their 200-nanosecond passage through a distant test mass. They have determined that the impulse beam remains collimated and undiminished in intensity over distances as great as 200 kilometers. Intervening grounded metal sheets and brick walls have no effect on shielding the beam. Tests carried out at a restricted government facility near Moscow have further demonstrated that when the discharge voltage is cranked up toward 10 million volts, the impulses are violent enough to dent a half-inch-thick steel plate or punch a four-inch hole through a concrete block.[12]

The effect that is most significant from the standpoint of space propulsion is that the Podkletnov beam generator produces no recoil force when it fires, the momentum of its electron discharge being entirely absorbed by its anode target electrode. Thus, the propulsion effects produced by the exiting gravity impulse are won without paying the price of any backward recoil. Let's suppose we are in a spaceship whose nose is filled with a lead mass. If we stand at the back of the ship and repeatedly fire a shotgun into the front of the ship, we will discover that we will not have moved at all. Our ship remains stationary because the forward impulse exerted by the absorbed buckshot is exactly canceled out by the backward recoil produced by the barrel of our firing gun. On the other hand, if we discharge a series of pulses from Podkletnov's gravity beam projector through the lead mass at the front of the ship, we will find that our ship will gradually accelerate. It is accelerated by the gravitational field gradients of the fleeting impulses as they momentarily pass through our ship's lead bulkhead. That is, our ship will be propelled forward by *gradient* drive.

Now, what will happen if we keep firing our gravity beam projector for an indefinitely long period of time? Theoretically, we should accelerate to faster and faster velocities. Podkletnov's research team was able to measure the speed of the impulses and has determined that their speed surpassed the resolution limit of their experimental setup, which indicated they were traveling faster than 64 times the speed of light![13] Consequently, since the gravity impulses are themselves superluminal, in principle they should cause our spaceship ultimately to accelerate to a superluminal velocity. In light of recent research, then, it seems entirely plausible that one day our own civilization will build spaceships that will be capable of superluminal speeds.*

Subquantum kinetics accounts for the superluminal results of this experiment. It proposes that the shock front produced by a sudden electrical discharge induces an *ether wind* that moves along with the shock at a superluminal velocity. Note that, unlike the speed of light limit, there is no known speed limit applicable to the subquantum level—that is, applicable to the ether. Consequently, this supposition does not violate any known laws of physics. Relative to its local ether

---

*I have worked with researcher Alexis Guy Obolensky, who has produced similar superluminal pulses by radiating electric field shocks from a dome-shaped electrode. Close to the dome we measured pulse speeds of up to 10 times the speed of light, which declined to luminal speeds as the distance from the dome increased. Subquantum kinetics had predicted that these impulses should exert a repulsive force on a distant pendulum much like those coming from Podkletnov's device. Preliminary experimental results have shown that this is indeed the case, although because the electron discharges that Obolensky's device produces are less powerful, the gravitational thrust produced by their shock fronts is also much less.

frame, the ether wind frame, the shock would propagate forward at the speed of light. But relative to the laboratory reference frame, its speed would be superluminal. Imagine a man running forward through a high-speed train. When his speed is considered relative to the train (the locally moving ether), he does not seem to be moving very fast, but when considered relative to the surrounding landscape (the galactic ether frame), he would be moving forward at an incredible speed. In the case of Podkletnov's device, his pulses are confined to a beam, and since the ether wind would be similarly confined, it would not diminish its velocity. Hence, the speed of his gravity impulses would remain constant with distance.*

The thought also presents itself that one day civilization might construct spaceports harboring immense gravity beam projectors whose beams would be large enough to envelop an entire spaceship and which would each be directed toward a particular star system. When placed in a propulsion beam, a ship would accelerate forward and quickly attain a superluminal speed. At the destination point, another propulsion beam would be turned on to decelerate the ship. Trips over a distance of several hundred light-years, for example, might be made in a matter of days. Perhaps a network of such spaceports is already in operation in our Galaxy.

Spaceflight Navigation

If extraterrestrial civilizations have developed superluminal spaceflight, they would need a means for charting their course through the Galaxy. One might consider, then, whether the primary function of the pulsar network is for use in space navigation. It might function something like our own Global Positioning System. The GPS is an array of radio signal–transmitting satellites that have been placed in geosynchronous orbit to ensure that they remain stationary above the Earth's surface. By triangulating their signals, cars, ships, aircraft, and hikers are able to accurately locate their position on the Earth's surface.

---

*In the case of Obolensky's experiment, on the other hand, because his pulses radiate outward from a dome electrode, the associated ether wind would fan out and diminish in velocity with increasing distance. Hence, his shock impulses similarly would decline in speed and approach luminal velocity, just as we observed.

NASA scientists used the pulsar network for navigation purposes when they sent out their ET communication message on board the Pioneer 10 spacecraft. As described earlier, an alien civilization coming across Pioneer 10 and its space plaque could locate the craft's Earth-based point of origin by identifying the particular pulsars designated on the plaque and using them to triangulate the Sun's location. In a similar fashion, by locking on to the signals of several pulsars, a space-ship could accurately locate its *instantaneous position* in the Galaxy. Equally important, the ship would be able to determine its *instantaneous velocity* by taking note of Doppler compression or expansion of the pulsar signals. Just as the pitch of a horn is Doppler shifted, increased or decreased depending upon the relative velocity of the listener, similarly the pulsation rate of a pulsar signal would be increased or decreased from its galactic rest frame pulsation rate depending upon the spacecraft's relative velocity.

For example, let us suppose that we are onboard a spaceship bound for a star near the Crab Nebula, Taurus constellation. Suppose that before we start out we target the Crab Nebula pulsar with our

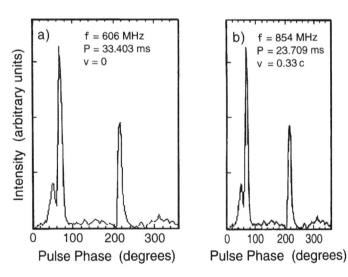

Figure 18. Pulse periods and pulse profiles for the Crab Nebula pulsar as seen from two reference frames: (a) the solar reference frame (v = 0) observed at a frequency of 606 MHz and (b) the in-flight reference frame (v = 0.33c) observed at a frequency of 854 MHz (left profile adapted from Moffet and Hankins, Astrophysical Journal, *figure 2).*

onboard radio telescope and, observing it at a radio frequency of 606 megahertz, we find it to have a pulsation cycle period of 33.403 milliseconds. Its time-averaged pulse profile would look as shown in figure 18a.* Now suppose that our spacecraft accelerates toward the Crab Nebula and reaches a velocity of 33 percent of light speed (0.33c). Upon reobserving this navigation beacon, we would find that the Crab pulsar now is pulsing 41 percent faster with a period of 23.709 milliseconds and would appear as shown in figure 18b. Also, the Doppler shift effect resulting from our velocity toward the pulsar would produce an apparent 41 percent increase in the pulsar's radio wave frequencies in direct proportion to the shortening of the pulsar's period. The radio frequencies that had been forming the pulsar's 606 megahertz pulse profile when we were at rest would now be observed by us to be blueshifted up to 854 megahertz. So by observing both the location and the pulse rate of pulsars, it should be possible to judge our direction and speed of travel.

However, to accurately judge our speed relative to the galactic rest frame, we would need to make a time correction. That is, since we are traveling at 33 percent of the speed of light, our onboard atomic clocks would be ticking about 6 percent slower due to the relativistic clock retardation effect. However, because we would not know our exact velocity, we would not know how much our clocks had slowed or how much of a correction to make to our pulse period observations.[†] Moreover, since we would not be able to accurately interpret the pulsar periods, we would not be able to accurately determine our velocity. Fortunately, there would be a way out of this dilemma. The pulsar beacons have been designed so that their periods gradually increase over time. By knowing the normal slow-down rate for a given pulsar in the galaxy rest frame and by reobserving that pulsar's slow-down rate from our spaceship, we can accurately determine the degree to

---

*It would probably be more realistic to assume that the space travelers observed the pulsar at a higher frequency in the gigahertz range, as the radio telescope dish for observing at this 606 MHz radio frequency would be rather bulky to take along on their trip. However, this example is suitable for the purpose of illustration.

†Measuring just the change in pulsar period will not be sufficient to allow a space traveler to accurately determine his velocity, because his measurement of the pulsar period is affected not only by his velocity but by the retardation of his clock as well. He could unambiguously determine his velocity from the pulsar period by using an independent method for determining his clock retardation, such as by measuring the pulsar period derivative.

which our clocks have retarded. For example, when our spaceship was initially at rest, we would have found that the Crab pulsar's period was increasing at the rate of about 1.088 millionths of a pulse period per day. While traveling at 0.33c, this period change would now seem to us to take place 5.605 percent slower, at the rate of about 1.030 millionths of a pulse period per day. Knowing this, we could now correct the pulsar period that we had observed, shortening it by 5.605 percent, from 23.709 milliseconds to 22.380 milliseconds. We could then realize that we are actually traveling at 0.33c in the galactic reference frame, 12 percent faster than what we thought our speed was before we made this correction.

The pulsar network, then, *is ideal for space navigation.* Being easily distinguished from one another, pulsars provide a means whereby a spacecraft may accurately determine its location through triangulation. They also allow accurate determination of velocity from the Doppler shift of the pulsar periods and of clock retardation from the change in the pulsar period derivatives. In addition, their broad spectrum makes these beacons ideally suited for high-speed interstellar flight. It allows their signals to be detected by narrow-band radio receivers onboard spacecraft traveling at almost any speed, regardless of the degree of Doppler frequency shifting.

Superluminal Communication

If one is not prepared to accept the possibility that civilizations might cross such vast stretches of interstellar space at superluminal speeds, one might still consider the possibility that they could share information among themselves at superluminal speeds. Is such communication possible? Maybe it is. In 1991, Thomas Ishii and George Giakos reported that they had transmitted microwaves at faster-than-light speeds.[14,15] Shortly afterward, in 1992, Enders and Nimtz, physicists at the University of Cologne in Germany, described transmitting microwaves through an undersized waveguide at superluminal velocity.[16] This work became more widely known after 1995, when this group succeeded in transmitting Mozart's 40th symphony through an undersized 11-centimeter-long waveguide at a speed 4.7 times faster than that of light.[17] Other physicists have been exploring the possibility of using quantum-entangled photon pairs as a way of sending information virtually instantaneously from one point in space to another.

Earlier researchers had explored a far less sophisticated technique that could transmit messages over great distances by means of longitudinal wave shock fronts. The physicist T. Townsend Brown, for example, had developed a communication device that generated its signals by repeatedly charging a capacitor to a high voltage and abruptly discharging it through a spark gap. The resulting energy shock fronts so produced were received by an electrified capacitor bridge that registered these waves as voltage transients read by means of a Brush chart recorder (fig. 19). His original bridge used titanium oxide capacitors. An investigator from the Office of Naval Research who witnessed a test of this device in 1952 reported that signals were successfully transmitted to a receiver located in an adjoining room within an electrically grounded metal shield.[18]

In a September 1953 patent disclosure, Brown described another version of this communication device that used massive electrically conductive spheres for both the transmitter and the receiver antennae and which was designed to convey audio frequency signals.[19] An article printed in *Interavia* magazine in 1956 indicated that Brown

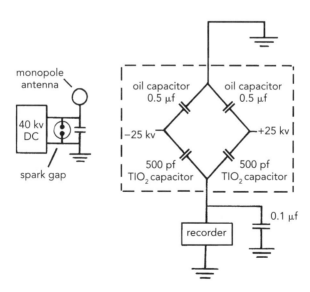

Figure 19. The electrogravitic communication transmitter (left) and receiver (right) developed by T. Townsend Brown.

suspected the waves being transmitted were not limited by the speed of light, although at the time he had no definitive proof.[20] Since he had determined that his capacitor bridge was able to detect gravitational disturbances, he concluded that the signals he was conveying must be gravitic, rather than electromagnetic. He reasoned that the waves were the gravitational homologues of light waves, which, for lack of a better word, he called "quasi-light." In further support of his gravitic hypothesis, he found that he obtained even better reception if he replaced the two titanium oxide capacitors in his bridge circuit with ceramic capacitors that had both a high mass density and a high dielectric constant.

Brown's transmitter bore a strong resemblance to the monopolar radio frequency generators built by the scientist and inventor Nikola Tesla in the late nineteenth and early twentieth centuries. These similarly produced sawtooth pulses and were radiated from a high-voltage antenna terminating in a metal sphere. Also, like Tesla's waves, Brown's pulses penetrated through Faraday cage shields. Like Tesla before him, Brown theorized that the waves he was producing were non-Hertzian. In other words, conventional Hertzian electromagnetic waves are produced when electrical charges oscillate from side to side—for example along an antenna dipole—and as a result their waves consist of forces that are oriented transverse to the direction of wave travel. Brown's waves, on the other hand, were produced by an oscillatory rise and fall of electrical charge on a monopole antenna, and thus consisted of field gradients that were oriented primarily *longitudinally* to the direction of wave travel. Whereas Hertzian waves travel at the speed of light in free space (their round-trip average velocity), longitudinal waves need not be similarly limited.[21]

The Podkletnov gravity impulse beam also operates by releasing shock discharges, and as mentioned earlier, researchers have measured its ability to transmit impulses at more than 64 times the speed of light. Moreover, since the beam has been shown to maintain its intensity over long distances, it has the potential for serving as a true long-range superluminal communication device, one that could send messages between star systems.

In summary, existing experimental evidence suggests that faster-than-light communication is definitely possible in the laboratory. Assuming that communicators can be built to transmit superluminal signals over interstellar distances, message time lags between civilizations would be minimal, thereby making possible an integrated galactic Internet. Could the Hertzian electromagnetic emissions from pulsars contain a non-Hertzian superluminal component, as yet unidentified, that permits such rapid communication? Or would such communications be transmitted from entirely different beacon arrays?

FOUR

THE GALACTIC
IMPERATIVE

＊•＊•＊

Message in the Stars

Earlier we saw that Sagitta, the Celestial Arrow constellation, closely indicates the Galaxy's northern one-radian point, the point along the galactic equator that marks out one radian of arc from the Galactic center. A close study of constellation lore reveals that Sagitta is part of a larger constellation cipher, one that includes the southern constellations of Centaurus and Crucis as well as the constellations of the zodiac and their astrological lore. As explained in my books *Genesis of the Cosmos* and *Earth Under Fire*, these ancient asterisms and their associated lore use the language of metaphor to convey that our Galactic core became explosively active and that cosmic rays and radiation from this extended outburst began to shower our solar system around 16,000 years ago, bringing about a global climatic disaster.[1,2] By designating the Galaxy's northern one-radian point, this constellation message conveys the idea of how the energy released from this Galactic center explosion had traveled radially outward through the Galaxy to ultimately impact the solar system.

Let us briefly summarize the conclusions reached in these books. The twelve zodiac signs and their ancient astrological lore are found to encode a physics that describes the spontaneous explosive creation of matter and energy.[3] Furthermore, using constellation pointers, the zodiac indicates the Galactic center as the location of this energetic outburst. That is, according to star lore mythology, Sagittarius (the

Archer) is aiming his arrow tip (Gamma Sagittarii) at the Heart of the Scorpion, which is specifically represented by Alpha Scorpii, a bright red supergiant star. Regressing these star positions back in time, we find that the Archer's arrow shaft would have been precisely aligned with the Scorpion's heart around 13,865 B.C.E. With this trajectory sighting, Sagittarius's arrow indicates the location of the Galactic core to within 0.35° of arc (fig. 20). Knowing too that Sagittarius and Scorpio together represent archetypal symbols for the first emergence of this created matter and energy, the arrow indication along with its encoded date would record a time when a Galactic center outburst became apparent to Earth observers—the date when this energy volley began passing through the solar system.

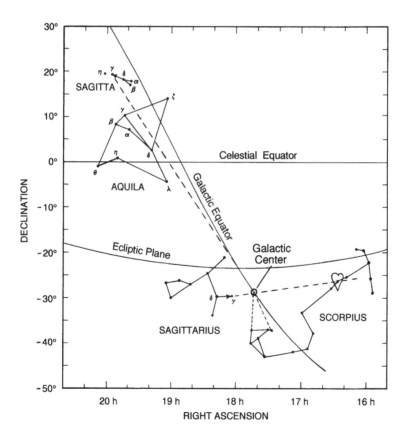

Figure 20. A sky map showing the locations of the constellations of Sagittarius, Scorpius, Sagitta, and Aquila (from LaViolette, Genesis of the Cosmos, *figure 9.20).*

Sagitta, the Celestial Arrow, which is portrayed flying away from the Galactic center with its point close to the northern galactic one-radian point, is Sagittarius's arrow, which has been explosively ejected from the Heart of the Scorpion—in other words, from the exploding Galactic center. Considering that the arrow tip, Gamma Sagittae, is the closest constellation star to the northern galactic one-radian point, the Celestial Arrow symbolically represents an energetic volley of cosmic ray particles and electromagnetic radiation that originated from the Galactic center and traveled radially outward the 23,000-light-year distance from the Galactic center to our solar system. Recall that an arc beginning at the Galactic center and extending outward in the plane of the sky over an angle of one radian has a length equal to the distance from the Galactic center to the Earth.

This raises the question that, by designating this same one-radian concept, perhaps the millisecond pulsar beacons PSR 1937+21 and PSR 1957+20 are referring to this same Galactic core explosion event. Indeed, if an astronomical phenomenon of this sort had occurred, it would indeed be the "talk of the Galaxy," the phenomenon that extraterrestrial communications would most likely be discussing, since it is something that *all civilizations in the Milky Way have experienced in common.*

Galactic Superwaves

Earth Under Fire presents astronomical and geological evidence indicating that a protracted global climatic disaster did occur shortly after this 13,865-year-B.C.E. date. These findings suggest that about every 13,000 to 26,000 years, the immense radiant mass at the core of our Galaxy enters an explosive phase that lasts up to several thousand years, during which it produces a fierce wind of cosmic ray particles and electromagnetic radiation. This forms an expanding spherical shell of radiation called a *superwave* that travels radially outward from the Galactic nucleus at close to the speed of light and penetrates entirely through and beyond the surrounding spiral arm disk (see fig. 21).[4,5] Less intense superwaves could recur more frequently, perhaps every 500 years.

Since the mid-1960s, astronomers have known that the massive cores of galaxies periodically enter a quasarlike active phase during which they intensely radiate cosmic ray particles and electromagnetic emissions. In 1983, when I first proposed the superwave concept, astronomers were assuming that Galactic core explosions occur much less frequently, about every hundred million years, and that their expelled particles were trapped within a few hundred light-years of the Galaxy's nucleus, although in some cases, where long-range propagation was evident, they assumed the particles were confined to narrow beams. I demonstrated that this was incorrect, that observational evidence indicates outbursts recur far more frequently and that their expelled cosmic rays travel outward at close to the speed of light to penetrate the entire Galaxy. Radio-telescope observations of the Galactic center recently confirmed this model of long-range radial propagation with the discovery that synchrotron radiation from our Galactic core is mainly *circularly* polarized.[6] Circularly polarized synchrotron radiation is produced only when cosmic rays travel *toward the observer at close to the speed of light,* following a spiral trajectory. Hence, circularly polarized radio emission coming from the Galactic core would indicate that the cosmic rays producing this radiation would be traveling radially away from the Galactic center.

As a superwave sweeps through the Galaxy, it has a major effect on the numerous sunlike star systems it encounters. Normally, the continual outward sweeping action of a star's ion wind is able to keep

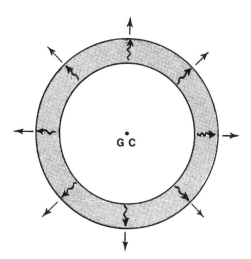

Figure 21. A schematic representation of a galactic superwave. Cosmic ray particles and electromagnetic radiation move radially outward from the Galactic center in the form of an expanding shell (from LaViolette, Earth Under Fire, figure 3.1).

its immediate environment clear of interstellar dust. However, at times when a reasonably intense superwave cosmic ray volley is passing by and is sustained over a period of many centuries, dust and gas residing near a star would be pushed forward and would acquire sufficient momentum to invade the star's planetary environment. As this material penetrated inward and fell under the influence of the star's gravitational pull, much of it ultimately would crash onto the star's surface with great force. The released energy would add to that which the star generates internally, thereby causing its luminosity to increase. Energy would also be supplied to the star's surface from its newly acquired dust shroud. This would intercept a considerable fraction of the star's outgoing visible light and return a portion back toward the star. These effects together would energize the star's surface and induce it to engage in violent flaring activity. This enhanced flaring could expose any planets in the star's solar system to high levels of radiation that would be hazardous for any life they might support.

The presence of this interplanetary dust would also substantially alter the climate of such planets, less radiation coming directly from the star and more coming indirectly as scattered or reradiated light. This indirect component would have warmed a planet relatively uniformly as though it were in a hothouse. This warming would have had its greatest effect in the polar regions of a planet, since these normally receive little direct radiation due to the low angle of incidence there. Thus, a planet's polar regions would have tended to become warmer than normal.

In addition, since the dust would absorb a portion of the star's light and reradiate it as infrared, the star's spectrum would have become reddened, thereby changing the amount of energy that would penetrate to the surface of the planet through its atmosphere. Together with the enhanced radiation output from the star, these effects could induce either climatic warmings or climatic coolings, the net result being dependent upon the magnitude of these various radiation processes and the radiation absorption properties of the planet's atmosphere. Calculations performed for our own planet demonstrate that the climatic impact would be quite sizable, sufficient to either initiate an ice age or end an existing glacial period, depending on the circumstances.[7]

Astronomical and geological evidence suggests that the most recent major superwave passed our solar system toward the end of the last ice age.[8] This is corroborated by ice-core data from Byrd Station, Antarctica, which indicates that the intensity of galactic cosmic ray radiation striking

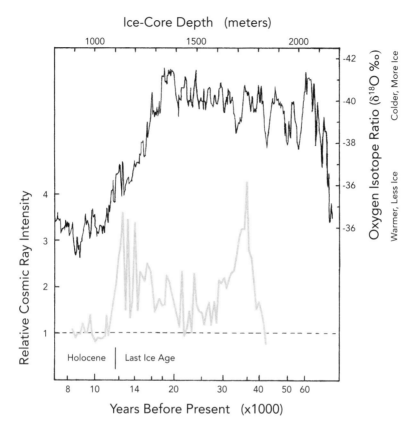

Figure 22. Lower profile: Cosmic ray intensity impacting the solar system (0–40,000 years B.P.) normalized to present levels. [Based on the Byrd Station ice core ^{10}Be concentration data of Beer et al. (Nuc. Instrum. Meth. Phys. Res., p. 204, and The Last Deglaciation, p. 145), adjusted for changes in ice accumulation rate and solar wind screening.] Upper profile: The ice core's oxygen isotope ratio, an indicator of ambient temperature and glacial ice sheet size (courtesy of W. Dansgaard).

the Earth's atmosphere rose to moderate levels around 16,000 years B.P., where B.P. signifies years before present as referenced from the year 1950 C.E. Cosmic ray intensity reached a peak around 14,100±200 years B.P., again around 13,250±200 years B.P., and climaxed a third time around 12,650±200 years B.P.; see lower profile in figure 22.[9] The last phase, the most prolonged of the three, lasted for about 1,500 years, all three peaks spanning a period of about 3,000 years. This period of elevated cosmic ray intensity coincided with the climatic warming trend that ultimately ended the ice age (see upper isotope profile in fig. 22). This graph, which plots cosmic ray intensity normalized to present-day intensities, was constructed from measurements of the concentration of beryllium-10 (^{10}Be) present in polar ice.*

Other evidence found in the Byrd Station ice-core record indicates that immediately before the ending of the last ice age, large quantities of acids had entered the Earth's atmosphere, more than at any time in the last 50,000 years.[10] This has been called the Main Event. This influx of hydrochloric and hydrofluoric acid began around 15,830 years B.P. and ended around 15,735 years B.P., rising and falling in amount in a regular fashion over a period of about one century. In 2005, I showed that the period between these acid peaks approximates the eleven-year solar cycle, which indicates that these acids and their associated dust must have a cosmic origin.[11] Satellite observations have shown that the solar cycle dramatically modulates the rate at which interstellar dust currently enters the inner part of the solar system. I had proposed that the same thing was happening during this ice-age event except that at that time the nebular influx was many orders of magnitude greater than present-day rates. Based on the acid concentrations found in Antarctic ice, I was able to estimate the interstellar-dust concentrations that would have been present in the solar system at the time this material was being deposited and was able to infer that this would have been sufficiently dense to absorb 18 percent of the direct solar beam and reradiate it in the infrared. The results showed they would have been climatically significant, in effect validating the model estimates I had made more than twenty years earlier. During this event, global temperatures initially decreased by about one degree Centigrade, but thereafter they progressively warmed over a period of several thousand years, eventually reaching interglacial temperatures.[12] The long-term warming would have been due both to the warming effect of the interplanetary hothouse effect and the reddening of the Sun's spectrum.

The increased luminosity of the Sun's photosphere due to the effect of this invading dust on the Sun and its tendency to aggravate solar flare activity would also have contributed to this global warming effect. In

---

*^{10}Be is a medium half-life element produced by the collision of high-energy cosmic rays with nitrogen in our atmosphere. Consequently, by determining the rate at which ^{10}Be was being produced in the atmosphere and subsequently accumulating on the Earth's surface, one may determine the intensity with which galactic cosmic ray radiation was impacting the Earth. Furthermore, by estimating solar flare activity levels in the past, one may determine the degree to which the solar wind was impeding the entry of galactic cosmic rays, making it possible to infer the unattenuated intensity at which the incident cosmic ray barrage was impacting the solar system.

fact, lunar-rock data show that the period at the end of the last ice age, between 16,000 and 10,000 years B.P., was distinguished by intense solar flaring activity.[13–15] Geologic records indicate that the warming of the Earth's ice-age climate dramatically accelerated beginning around 14,700 years ago, when temperatures in high-latitude regions reached near present-day values for almost two thousand years.[16] This was accompanied by a period of rapid ice-sheet melting and continental flooding. There is no known terrestrial mechanism that can account for such a rapid warming of the whole planet! However, such a climatic shift would have been an expected consequence of an activated Sun and a dust-congested interplanetary medium.

Excess radiocarbon levels found in ice-age sediments indicate that solar activity reached a peak around 12,750 years ago,[17] a time that coincides with the worst episode of mass animal extinction to occur in millions of years.[18] Together with the lunar rock evidence, this suggests that the Earth and Moon were engulfed by a highly intense solar coronal mass ejection that exposed the Earth's surface to lethal radiation levels. This prolonged disaster may have spawned myths and legends describing celestial phenomena wherein a previously darkened Sun violently erupts to singe the Earth and trigger the release of vast deluges that wash over the land, events that are said to have nearly extinguished the human race.

It is interesting that terrestrial constellation lore anticipated the date of the Main Event cosmic-dust incursion. I first announced the discovery of this zodiac lore date in 1995 with the publication of the book *Genesis of the Cosmos* and elaborated on it in 1997 with the publication of *Earth Under Fire*.[19,20] At that time I was a bit puzzled as to why constellation lore would be indicating a time before much of the action had taken place, the main climatic warming, continental flooding, and mass extinction events having occurred some one to three thousand years after that date. No unusual findings had then been reported that would correlate with that date except that climate had begun its long-term warming trend around that time. Hammer, Clausen, and Langway first published their discovery of the Main Event acid feature in 1997, and I first became aware of this in 2000. It was in the subsequent years that I discovered the event's cosmic origin through its solar cycle fingerprint. Here, then, is an example in which science has later come to verify important geological information hidden in myth lore that has been unknowingly handed down from generation to generation.

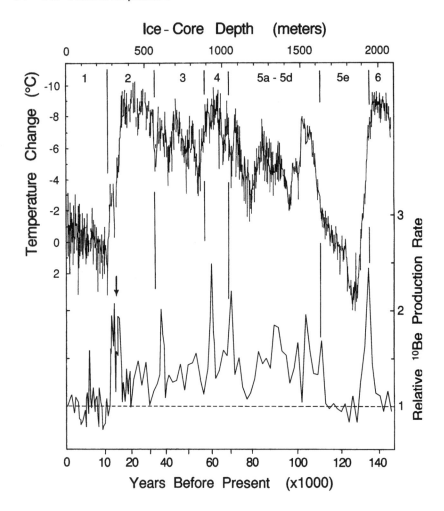

Figure 23. Lower profile: Cosmic ray intensity impacting the solar system (0–145 kyrs B.P.) normalized to present levels (based on the Vostok, Antarctica, ice core ^{10}Be concentration data of Raisbeck et al. [The Last Deglaciation, p. 130], adjusted for changes in ice accumulation rate and solar wind screening). Upper profile: Ambient air temperature, as indicated by the ice core's deuterium content (from Jouzel, Nature, p. 403).

The graph shown in figure 23, which plots data from the Vostok ice core, presents a longer-term view of the relative cosmic ray intensity that has been striking the solar system.[21] As in figure 22, the cosmic ray intensity plotted here was determined from measurements of ^{10}Be accumulation rate and from estimates of solar flare activity. The data indicates that several large galactic cosmic ray events occurred during the last ice age, which dates from 110,000 to 11,550 years B.P. Also, another large peak about 132,000 years B.P. is apparent at the end of the preceding ice age. The peaks at 14,150±200 years B.P. and at 12,600±200 years B.P. are seen to be the most recently occurring of the major cosmic ray events, although a very brief, small-scale event may have occurred around 5,350 years ago (fig. 23).

The cosmic ray peak dating from around 37,000 years B.P. has received a considerable amount of attention since it has been detected in several ice-core records. It is of particular interest since it coincides with the demise of Neanderthal man. The scientific community has reached a consensus that this peak indeed represents a period of enhanced cosmic ray intensity. However, the ice-core record contains other equally prominent peaks. Galactic superwaves can recur often enough to explain them, whereas nearby supernova explosions occur far too infrequently. In particular, there is independent astronomical data to corroborate the passage of a superwave during the period from 15,000 to 11,000 years B.P.[22,23]

The corresponding cosmic ray peaks at the end of the last ice age are not readily discerned from the raw ^{10}Be concentration ice-core data. Hence, their presence has until now eluded ice-core researchers and as a result they have received no discussion in the scientific literature. Their presence becomes apparent only when the raw data is properly adjusted to take account of past changes in ice accumulation rate and in the degree of solar wind screening.* That is, a rise in annual snow-fall increases the ice accumulation rate and yields greater ^{10}Be dilution, hence lower ^{10}Be concentrations. Also, greater solar wind screening more effectively deflects incoming galactic cosmic rays and results in lower ^{10}Be concentrations.

---

*In producing this profile, data values for the period 15,000 to 10,000 years B.P., which represent the intensity of galactic cosmic rays striking the Earth, were adjusted upward substantially so that the profile would accurately represent the cosmic ray intensity impacting the solar system. This upward adjustment is necessary since ^{14}C data indicates that solar flare activity was at least an order of magnitude greater during that period and hence that the solar wind would have had a substantially greater screening effect on the incident cosmic ray flux. No similar solar wind adjustments were made to earlier cosmic ray peaks due to insufficient data on solar activity that far back in time. By using a detailed ice-core climatic profile for the period 20,000 years B.P. to the present, it was possible to accurately date the Vostok core through correlation with the well-dated Summit, Greenland, ice-core profile. This, in turn, allowed the rates of ice accumulation and ^{10}Be production to be accurately determined. Details of the derivation of the profiles shown in figures 23 and 24 are described in my previous publications, *Galactic Superwaves and Their Impact on the Earth* and "Evidence for a global warming at the Termination I boundary and its possible extraterrestrial cause."

Although scientists have today amassed a considerable store of knowledge about Galactic core explosions seen happening in the nuclei of distant galaxies and have gathered evidence that such outbursts have also poured forth from the center of our own Galaxy, still the superwave phenomenon is quite elusive. The quiescent periods between outbursts are so extensive that any knowledge of the previous event would be long forgotten, surviving only in scattered myths that report some of the terrestrial consequences. The cosmic ray electron volley leaves little hint of its presence in the Galaxy once it has passed. The electrons themselves are invisible, their presence being evidenced only through the synchrotron radiation they emit. But since they beam their radiation away from the Galactic center, their emission is not easily discerned once the superwave has passed the observer. Moreover, due to their speed-of-light flight through the Galaxy, subsequent superwaves would arrive with little warning. It appears that some ancient civilization attempted to bridge this gap in time by sending us a time-capsule message with the purpose of warning future generations about the phenomenon. Indeed, were it not for this ancient message embedded within constellation lore, we might not have known today about the superwave's past attack.

Could the pulsar network be referring to this same galactic cosmic ray phenomenon? Just like the constellation cipher in ancient terrestrial star lore, the pulsar signals beamed in our direction explicitly refer to the northern galactic one-radian point as seen from our particular galactic location. In so doing, they imply not only that they have knowledge of the one-radian geometrical concept, but also that they know the location of the Galactic center and that events happening at this central location affect our solar system.

It should not be entirely a surprise to find that this last superwave is the topic of interstellar conversation. Indeed, all worlds in the Galaxy must endure the consequences of a superwave's passage. In choosing a subject for interstellar communication, it makes sense that another civilization would select a phenomenon that we and other civilizations in our direction would have similarly experienced. Just as a volcanic eruption or major earthquake becomes the subject for widespread discussion on the evening news, so too we can expect that the most recent superwave would be a prime topic for conversation in interstellar ETI transmissions.

There are also altruistic motives that would encourage interstellar conversation about superwaves. The more advanced civilizations in the Galaxy might warn others who are less informed about past occurrences of this phenomenon. Or they might even share information about when the next superwave would arrive, perhaps conveying information about this through the galactic network by transmitting signals that travel far faster than light. They might also explain ways in which a civilization could protect itself from a superwave attack—for example, by devising ways of deflecting the trajectories of approaching cosmic rays.

Given that unique pulsar beacons have sky positions so close to that of the key constellation star Gamma Sagittae, the speculation comes to mind that the Sagitta constellation and the zodiac constellation lore with its encoded time capsule cipher might all be of extraterrestrial origin. This "ancient astronaut" theory suggests that shortly after the decimating passage of the last major superwave, our planet was visited by advanced beings from a nearby star system whose civilization was a communicating member of the pulsar network. They could conceivably have devised a constellation lore for Earth's survivors whose purpose was to inform future generations about this catastrophe and about the superwave that caused it. In so doing, they could have arranged that this time-capsule message and the galactic pulsar network message would cross-reference one another.

In fact, there are indications that whoever designed the zodiac constellation lore possessed *a very advanced knowledge of science.* For example, when we consider the trajectory of the Archer's arrow sighting toward the Heart of the Scorpion, consider also the location of the Sagitta arrow at its position close to the northern galactic one-radian point, and consider the location of the Southern Cross marker, Crucis, which accurately indicates the southern galactic one-radian point, we must conclude that whoever configured this system of asterisms knew the location of the Galactic center to within a few tenths of a degree of arc. Modern scientists have bettered this accuracy only in recent decades by using large-dish radio telescopes. Also, the zodiac symbolically designates a temperature gradient in space that extends progressively from a hot pole in the direction of Leo to a cool pole in the direction of Aquarius. Modern scientists confirmed this just recently when they discovered a similar dipole temperature anisotropy in the 3 Kelvin cosmic microwave background radiation.[24] They accomplished this through the use of stratospheric flight, sophisticated electronic circuitry, and cryogenically cooled microwave detectors.

In addition, a considerable amount of scientific knowledge was apparently involved in crafting the Virgo constellation and its lore. Myths describe her as seeding stars throughout the universe. The actual constellation shows her pointing to the center of the Virgo supercluster with her right hand while scattering stars out along the supercluster equator with her left hand.[25] The Virgo supercluster is the largest collection of galaxies in our part of the universe, our own galactic cluster being an outlying member. To be aware of the unique qualities of this location, an ancient civilization would have needed to have access to an optical telescope capable of resolving the images of distant galaxies, a telescope having an aperture of at least six inches. Moreover, their science would have had to be sufficiently advanced to have a knowledge of galactic redshift spectra, thereby allowing them to understand that the images they were seeing were distant galaxies and not local gas clouds.

Ancient writings do refer to sightings of advanced airborne craft and of past contact with extraterrestrial beings.[26,27] Viewed within a historical context that acknowledges the existence of such contact, the notion that ancient terrestrial star lore might interface with a message symbolically conveyed through a network of interstellar communication beacons does not seem all that far-fetched.

FIVE

SUPERWAVE WARNING BEACONS

The Crab and Vela
Supernova Remnants

There is another part of the pulsar message that makes even more explicit reference to the galactic cosmic ray volley that passed the Earth at the end of the last ice age. It is conveyed by the Crab and Vela pulsars, two very unusual pulsars that are not found near the one-radian points. One important characteristic that distinguishes them from most others is that they are associated with supernova remnants. As we will see, this supernova connection is crucial to understanding their superwave message.

The Crab pulsar (PSR 0531+21) is associated with the Crab Nebula supernova remnant (see fig. 24), which lies near the outer edge of the Milky Way in the constellation of Taurus, the Bull. Our solar system lies about 23,000 light-years from the Galactic center, and the Crab Nebula is situated 6,585 light-years farther out, the Galaxy having a radius of roughly 35,000 light-years. The Crab supernova was observed by Chinese astronomers in 1054 C.E., making it one of the few bright supernovae to occur in the last millennium.

The Vela pulsar (PSR 0833–45) lies along the line of sight of the Vela supernova remnant (see fig. 25), which is situated just 815±100 light-years away in the constellation of Vela, the Sail. The date of the Vela supernova is not known as well as that of the Crab, as it occurred so long ago. But astronomers often give its age to be in the range of

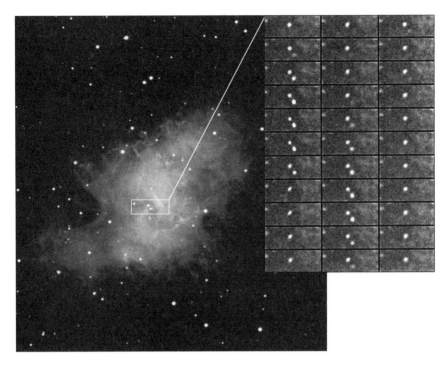

Figure 24. Image of the Crab Nebula with a time sequence inset showing the Crab pulsar's optical pulses over the course of one cycle. Each of the 33 time slices represents an interval of approximately one millisecond. The pulsar's brighter main pulse is visible in the first column; its less intense, broader interpulse can be seen in the second column. Kitt Peak National Observatory (courtesy of N. Sharp, AURA, NOAO, and NSF; Sharp, P.A.S.P., figure 9).

10,000 to 12,000 years.[1] Sumerian star lore tells of a "giant star" (i.e., supernova) that once appeared in the constellation of Vela, which it identifies with the god Ea, the Lord of the Waters.[2] The Sumerians regarded Ea as the god of Wisdom, who taught humanity the arts of civilization. More significant, their flood myth holds that Ea was the god who warned Ziusudra (Noah) of the impending flood disaster that was about to drown the world. The geologic record indicates several times at the end of the last ice age when the ice sheets were rapidly melting and flooding the land, but the most significant of these events is the one that occurred 10,750±100 B.C.E. which marks the time of the terminal Pleistocene mass extinction. A study of animal remains indicates that this disaster was indeed associated with cataclysmic floods occurring on a global scale.[3] If this is the global flood that Sumerian legend

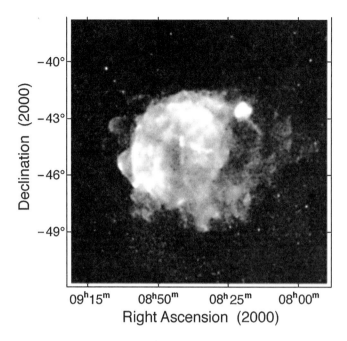

Figure 25. X-ray image of the Vela supernova remnant imaged with the ROSAT X-ray satellite (courtesy of B. Aschenbach, ROSAT, and Nature*).*

describes, then the day's of Ea and the associated Vela supernova might be dated back to that same period, circa 10,750 B.C.E., which corresponds reasonably well with accepted estimates for the age of this remnant. Since the Vela supernova occurred relatively close to the solar system, it would have been quite bright and could have made quite an impression on ice age cultures.

Pulsars Are Not Made in Supernova Explosions

Conventional theory assumes that pulsars are spinning neutron stars that have been born out of the crushing pressures of a supernova explosion. The vast majority, 97 percent, of the more than 1,533 known radio pulsars exhibit no supernova remnant associations, and astronomers presume this is because they are so old that their expanding supernova remnants have long since dissipated. But if the neutron star theory is correct, neutron stars should be found within all of the 231 known supernova remnants, and even if their synchrotron beams are not aimed in our direction, they should make themselves evident from the X-rays their hot surfaces would be radiating in all directions. However, as of the time of writing, *only 50 supernova remnants* have been found to be associated with pulsars. As for the other remnants, X-ray telescope observations have turned up little evidence of compact energetic X-ray sources. Thus, we are left with only 50 detections out of 231.

The paucity of pulsar detections in the observable set of remnants has been disturbing to pulsar astronomers. For example, in their 1986 review paper, the pulsar astronomers Joseph Taylor and Dan Stinebring wonder:

> Why, among the ~150 known galactic supernova remnants, have pulsars been found in only three? Even at a few hundred kilometers per second, pulsar velocities are much less than the several thousand kilometers per second typical of supernova ejecta. Therefore a pulsar created in a supernova event should remain well inside the supernova remnant as long as the remnant is visible, and it should be relatively easy to find. Efforts to find such pulsars have not been very successful.[4]

Among those pulsars that are found within the perimeter of a supernova remnant, there is still a question as to whether they in fact originated from the remnant's supernova explosion. The Vela pulsar is one example. It is located at ℓ = 263.55°, b = –2.79°, about 1.3° of arc from the center of the Vela supernova remnant's explosion center, which is located at ℓ = 263.9±0.2°, b = –1.8±0.2° (see fig. 26). Since the remnant is 5° in diameter, the pulsar is seen to reside within its perimeter. But estimates of the pulsar's distance place it twice as far away as the remnant—i.e., about 1,600 light-years from us.[5] This association, then, may be just a projection effect.

But even if we were to assume that the Vela pulsar lies the same distance from us as the Vela remnant (i.e., 815 light-years), the evidence still weighs against it having originated in the Vela supernova explosion. The pulsar then would have had to travel 18 light-years from the explosion center in the time that has elapsed since the Vela supernova explosion, meaning that it would have had to have traveled at a speed of about 440 kilometers per second. Instead, measurements of the pulsar's proper motion show that if it were at the same distance as this nebula, it would be moving *at only about one seventh of this speed* (62±4 km/s) and along a trajectory that is not aligned with this explo-

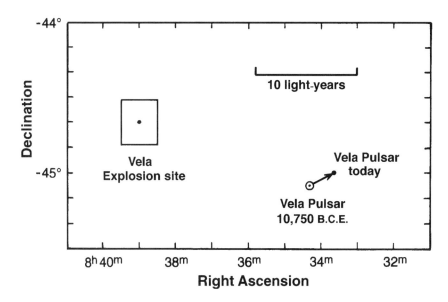

Figure 26. Location and trajectory of the Vela pulsar as compared with the location of the Vela supernova explosion site.

sion site.[6] Consequently, it is very unlikely that this pulsar originated from this supernova explosion. As a result, some have voiced doubts as to whether it is indeed associated with the Vela supernova remnant.[7]

The Crab pulsar, however, does show evidence that it emerged from its supernova explosion site (see fig. 27). As viewed in the plane of the sky, the Crab pulsar is currently located at $\ell = 184.56°$, $b = -5.78°$, about 0.35 light-year from the site of the Crab supernova explosion.[8] Knowing the pulsar's presently observed proper motion,[9] we can project back in time to find the position it would have had in 1054 c.e., the date of the supernova. We find that on this date the pulsar's location would have coincided with the supernova explosion center to within the error of measurement. Hence, the Crab pulsar may be the stellar core remnant of the Crab supernova progenitor star—a very dense stellar core that radiates an energetic wind of cosmic ray electrons and which may or may not be a neutron star. Considering that the Crab pulsar is one of the few cases in which there is clear evidence for emergence from a supernova explosion site and considering the general scarcity of pulsar–supernova remnant associations, we are left to doubt the spinning neutron star theory, which requires pulsars to be formed in supernova explosions.

The direction of the Crab pulsar's proper motion matches that of two nearby line-emitting filaments positioned at the forefront of the remnant shell and having the highest radial velocity toward us of all known nebula filaments. Since the pulsar is positioned in projection almost three times closer to the explosion center, we may surmise that, like these filaments, it is located about five light-years from the explosion center at the forefront of the Crab Nebula shell and is moving almost directly toward us at close to the shell's maximum radial expansion speed, the pulsar's direction of movement deviating from our line-of-sight direction by just two degrees of arc. The pulsar's peripheral position could explain why interstellar dispersion of its radio pulses is observed to be slowly increasing.[10] If it were instead located *inside* the remnant, its signal dispersion should be decreasing by 0.07 percent per year due to the progressive expansion and thinning of the nebula's plasma. But this is not seen to be the case.

Could the Crab and Vela pulsars be ETI beacons carefully placed to mark these remnants for us? If so, why are they being marked, and why these particular remnants? One clue comes from noting that both have unique positions relative to our solar system. For example, among young

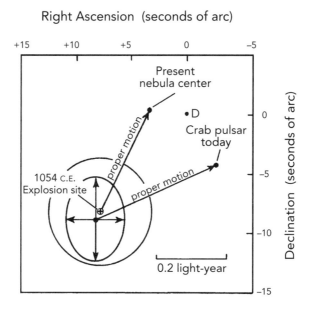

Figure 27. Location and trajectory of the Crab pulsar compared with the location of the Crab supernova explosion site.

supernova remnants, those having an age of less than 20,000 years, Vela is the *closest* to us, lying just 815 light-years away (figure 28). Its proximity is surpassed only by the North Polar Spur, a several-million-year-old remnant whose explosion center is located about 400 light-years away. The more distant Crab Nebula remnant has the distinction of being the *second closest* to our solar system among young remnants and of being the only remnant to have a sky position so close to the galactic anticenter. The galactic anticenter is the direction that lies exactly opposite the Galactic center direction, and whose position relative to background stars depends specifically on the viewer's location in the galactic disk. The center of the Crab Nebula is positioned at ℓ = 184.56°, b = −5.78°, just 7.4° from the anticenter position (179.94°, 0.05°); see figure 28.

The Crab Nebula also happens to be the most conspicuous supernova remnant in the sky. Unlike most other remnants, which require radio telescope observation to reveal their presence, the Crab Nebula is optically visible. It is the only one that can be easily seen with an amateur optical telescope. Is it just a coincidence that of all supernova remnants in the Galaxy, the two that happen to have unique placements relative to our solar system are among the few to be marked with pulsars, and as we will see, by two very unusual pulsars?

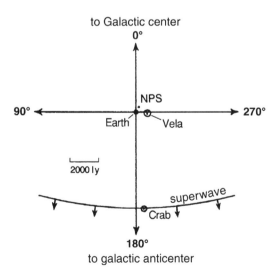

Figure 28. The positions of the Crab and Vela supernova explosions relative to the solar system.

Wave of Destruction

Supernova Triggering by Superwaves. To discover why these remnants might have been chosen to be marked with ETI communication beacons, we must look beyond the fact that they are uniquely placed relative to us. As we will see, both of these supernovae appear to have been triggered by the same galactic superwave that was responsible for the major cataclysm at the end of the last ice age. When we take account of the remnants' locations in the Galaxy and their ages, we find that their respective supernovae were asynchronously linked with one another, their detonation dates differing approximately by the amount of time it takes light to travel from the Vela site to the Crab site! It is, then, entirely reasonable that a superwave "event horizon," moving outward from the Galactic center at the speed of light and reaching the Earth around 14,130 years ago, first passed the relatively nearby Vela site, causing its supernova to occur, and then, after traveling some 6,300 light-years farther on, passed the Crab Nebula site, causing its supernova also to occur (figs. 28 and 29).[11,12]

A particularly intense superwave could have coaxed the progenitor stars of these various remnants to explode if these stars were hot, inherently unstable, and embedded in a dust-laden environment.[13–15] That is, upon their arrival, the superwave cosmic rays would have overpowered the star's stellar wind and pushed nearby dust in close to the star. As the star began to gravitationally draw this material onto its surface, the added kinetic energy would have provoked its energy output to rise abruptly, whereupon the star finally would have exploded. Alternatively, an advancing superwave might carry a steep gravitational field gradient that could induce frictional tidal forces sufficiently strong to energize the star and cause it to explode.

Because the Crab supernova was seen less than a thousand years ago, the superwave that triggered its explosion should still be in the remnant's vicinity, still impacting it with cosmic ray electrons. The ionized turbulent plasma within the remnant would be magnetically trapping these impacting particles, capturing them into tight-spiral orbits and causing them to continuously emit synchrotron radiation in all directions. This explains why the entire remnant is presently seen to be an intense emitter of synchrotron radiation at radio, optical, and X-ray wavelengths. At optical, X-ray, and gamma ray wavelengths, it is the *brightest* supernova remnant in the sky, and at radio wavelengths

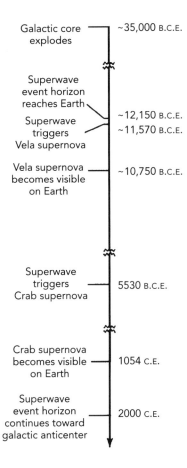

Figure 29. Event timeline for the sequential triggering of the Vela and Crab supernova explosions.

it is the second brightest. Astronomers concur that this emission is produced by cosmic ray electrons trapped in the remnant. The radio contour map shown in figure 30 illustrates the extent of the synchrotron emission from the nebula.

Being unaware of the superwave phenomenon, astrophysicists have concluded that the cosmic rays illuminating the Crab Nebula are being supplied from the Crab pulsar, which they presume lies inside the remnant. However, the majority of cosmic rays producing this emission should be coming from the passing superwave and *not from the pulsar.*

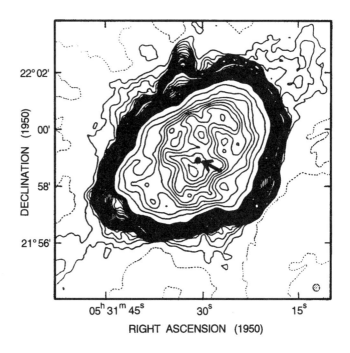

Figure 30. A radio contour map of the Crab Nebula showing the intensity of its synchrotron radio emission. The arrow indicates the position of the Crab pulsar (Reprinted courtesy of NRAO, Nature, and T. Velusamy, Nature, figure 2).

If the electrons producing the nebula's radio and optical emission were coming exclusively from the pulsar, then the synchrotron emission coming from the Crab Nebula and from its pulsar should be found to have very similar differential energy flux spectra. But they don't. In the radio and optical spectral regions, the Crab remnant and Crab pulsar spectra have entirely different slopes, indicating that they are produced by *different* populations of cosmic rays.[16-18] The Crab Nebula's synchrotron optical spectrum is negatively sloped (decreases in intensity with increasing frequency), whereas the Crab pulsar's optical spectrum is positively sloped. The Crab Nebula and the Crab pulsar both have synchrotron radio spectra that are negatively sloped, but the pulsar's radio spectrum is ten times steeper than that of the nebula, a slope of –2.5 as compared with –0.26 for the nebula. The Crab Nebula's radio spectrum makes a far better match to the nonthermal spectrum of the *galactic radio background radiation,* which has a slope of –0.4, a diffuse emission that astronomers agree is produced by galactic cosmic rays.* This diffuse synchrotron radiation comes from all parts of the sky, but is more concentrated toward the galactic plane and rises to its maximum intensity toward the Galactic center. More specifically, as I have shown in my dissertation, this radio background radiation is accurately modeled by a shell of superwave cosmic rays moving outward at the 12,150-B.C.E. superwave event horizon. A contour map of this radiation shows that the observed radio intensity varies with galactic longitude in the same way as would be expected if produced by these superwave cosmic rays.[19,20]

Among the category of young supernova remnants, there are two others whose, ages, locations, and distances from us are accurately known—the Cassiopeia A and Tycho remnants. Like the Crab remnant, we find that these are in the midst of the superwave particle barrage. The locations of these remnants in relation to the superwave may be seen in figure 31, which adopts a view looking down onto the galactic plane. The superwave event horizon does not plot out as a sphere, but rather as an ellipsoid with foci centered at the Galactic center and at the Earth. All points on this horizon are determined by the sum of: 1) the time required for superwave cosmic rays to travel from the Galactic center outward to that horizon and 2) the time required for synchrotron radiation from that horizon to travel back toward the Earth at the speed of light and be seen by us (see *Earth Under Fire* for an explanation).

The timing of the Tycho and Cassiopeia A supernova explosions suggests they were triggered by the same superwave front that triggered the Vela and Crab explosions. Taking into consideration the time required for the shell of superwave cosmic rays to cover the galactic radial distance from our solar system to the site of the Crab supernova (6,520 years), the time for light from the resulting supernova explosion to travel back to the Earth (6,585 years), and the date when this supernova was seen on Earth (1054 C.E.), we find that the event horizon triggering this explosion should have passed through the solar system around 14,000±60 years B.P. (see table 2).[21]

Performing the same calculation for the Cas A and Tycho remnants, we find dates of 14,670±500 years B.P. and 13,560±500 years B.P.,

---

*The galactic radio background should not be confused with the cosmic microwave background radiation, which instead has a thermal blackbody spectrum and is of intergalactic origin.

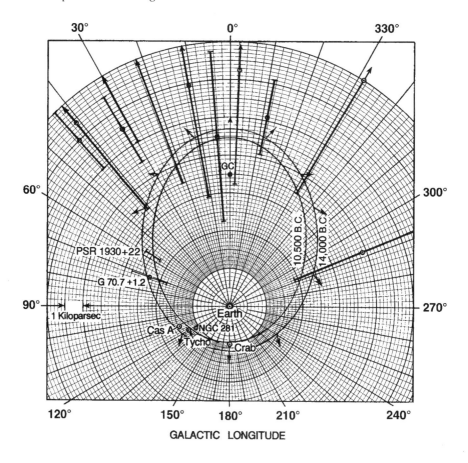

Figure 31. The positions of several young supernova remnants relative to event horizons for superwaves passing the solar system 12,500 and 16,000 years ago. Cosmic ray intensity would have peaked about midway between these two boundaries. The galactic plane lies in the plane of the paper. GC indicates the location of the Galactic center (P. LaViolette, Earth Under Fire, figure 10.4).

respectively. Averaging together the Crab, Cas A, and Tycho dates gives 14,080±600 years B.P. as the best estimate for the time when the most intense part of the galactic superwave would have passed the Earth. This compares favorably with the event horizon date of 13,620±2,000 years B.P. that would have been responsible for triggering the Vela supernova. This also correlates reasonably well with the 14,050-years-B.P. climax in galactic cosmic ray intensity registered in the Earth's polar ice-core records. Thus, we are left to conclude that *all of these supernovae were triggered by the same galactic superwave event hori-*

TABLE 2. INDICATION THAT A SINGLE SUPERWAVE TRIGGERED
FOUR SUPERNOVAE

Remnant	Distance from Us	Date of Visible Explosion	Earth Passage Date of Superwave
Crab	6,585 ± 30 l.y.	1054 C.E.	14,000 ± 60 yrs B.P.
Cas A	9,450 ± 300 l.y.	1658 C.E.	14,670 ± 500 yrs B.P.
Tycho	8,150 ± 300 l.y.	1572 C.E.	13,560 ± 500 yrs B.P.
Vela XYZ	820 ± 100 l.y.	10,750 ± 2000 B.C.E.	13,620 ± 2,000 yrs B.P.
	Average of Crab, Cas A, and Tycho:		14,080 ± 600 yrs B.P.

zon, whose Earth passage date is marked by elevated concentrations of cosmogenic beryllium in polar ice.

In view of the above, we are led to wonder whether the Crab and Vela pulsars are ETI beacons that were carefully positioned at the sites of the Crab and Vela supernovae to call our attention to the most recent galactic superwave event, one that was involved in triggering these explosions. Of the various supernovae that this superwave triggered, the remnant of the Vela explosion is closest of all to the Earth. Also, the remnant of the 1054-C.E. Crab explosion is closest of all to the galactic anticenter location. By specifically marking these particular remnants, then, a galactic collective would be conveying the idea of a disturbance moving away from the Galactic center at close to the speed of light.

Superwave Bow Shock Fronts. The Crab remnant is unique in that it is the only supernova remnant to show signs of motion in its interior. This is evident in the rapid sub-light-speed movement of *luminous wisps* seen near the Crab pulsar. Figure 32 shows a succession of optical images of the Crab Nebula made two months apart with the Hubble Space Telescope. These show substantial activity of the luminous region north of the pulsar (upward in the diagram), where luminous waves are observed to move northward at up to half the speed of light. This motion becomes naturally explained if superwave cosmic rays are traveling face-on toward the remnant (into the diagram) at close to the speed of light and are impacting a bubblelike bow shock front

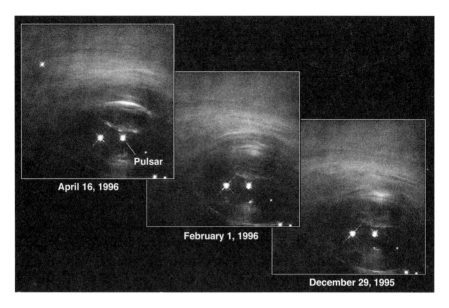

Figure 32. Successive images of the Crab Nebula made with the Hubble Space Telescope showing changes in the luminous wisps in the vicinity of the pulsar. The image has been rotated counterclockwise so that the nebula's long axis is disposed vertically (courtesy of J. Hester, P. Scowen, and NASA).

formed around the pulsar (recall fig. 16). As the superwave cosmic rays deflect around this obstacle, a large fraction will become magnetically trapped in the shock front and will emit optical synchrotron radiation, seen by us as the luminous wisps.

Since the radiant epicenter of the wisp movement is not centered on the pulsar, but rather about 0.15 light-year away, astronomers have concluded that this activity must be taking place some distance from the pulsar. Reluctant to relinquish their lighthouse model, they have attempted to attribute the phenomenon to a collimated beam of particles issuing from the pulsar and impacting the surrounding nebula. The beam is assumed to issue along the pulsar's stationary spin axis and to be inclined at a 60-degree angle to our line of sight. However, with such an orientation, the pulsar's synchrotron beam would not shine in our direction and we should be unable to see it pulse as the neutron star rotated. So by attempting to account for the wisp activity, the neutron star model runs into difficulty in explaining why we also see pulses coming from the Crab pulsar. Rather than being initiated by the pulsar, this wisp movement is most likely caused by the onslaught of superwave cosmic ray fronts arriving from the Galactic center and continuously bombarding the Crab Nebula and its pulsar beacon star face-on from our viewing direction.

This superwave volley also accounts for the broad plateau of diffuse X-ray emission that is positioned about 0.7 light-year northwest of the pulsar. Significantly, this arc-like plateau of synchrotron emission circumscribes the region with wisp activity and also coincides with the positions of the two luminous filaments having the highest radial velocity toward the Galactic center. This is exactly the layout we would expect for a face-on view of a superwave bow shock front.[22]

Several studies have been made of the orientation of the Crab Nebula's magnetic field.* These have shown that the magnetic field is oriented quite uniformly across the face of the nebula and in line with the its minor axis, hence perpendicular to the galactic plane.[23] The field maintains a similar orientation even in the vicinity of the pulsar, where it runs roughly parallel to the orientation of the luminous wisps. This suggests that the wisps are regions where the nebula's magnetic field is quite strong and thus better able to trap cosmic rays. Quite significantly, the nebula's magnetic field direction deviates by about 50° from the average field direction in the Crab pulsar that is responsible for producing its optical pulses. This finding led the astronomer William Cocke and his coworkers to conclude that the optical emission from the pulsar and its nebula environment are unrelated: "No obvious relation can be detected between the optical polarization of the pulsar and that of the Crab Nebula in its immediate environment."[24]

The lack of similarity between the pulsar and its immediate environment, in regard to both spectrum and polarization direction, provides strong evidence against theories suggesting that the pulsar is actively energizing its nebula, at least at radio and optical wavelengths, and in favor of the idea that most of the nebula's illumination is being externally supplied by a galactic superwave.

There also appears to be evidence of a superwave-induced bow shock front around the Cassiopeia A remnant. As seen in figure 33, most of the remnant's synchrotron radio emission comes from its western side (right side in the diagram), the side that faces the Galactic

---

*The orientation of the nebula's magnetic field has been inferred from the direction in which its optical synchrotron emission is polarized. Gyrating electrons always produce synchrotron emission whose plane of polarization is parallel to their gyration plane and perpendicular to the magnetic field direction.

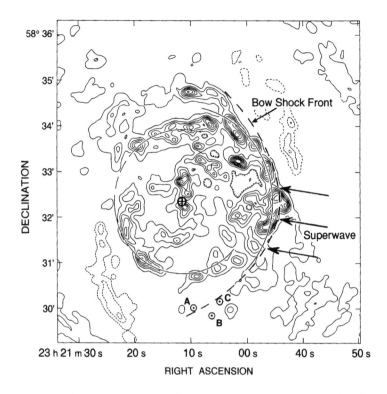

Figure 33. Radio contour map of Cassiopeia A made at a radio frequency of 2695 megahertz (adapted from Dickel and Greisen, Astronomy and Astrophysics, *figure 5). The central cross represents the position of the supernova explosion center. The dashed arc traces the bow shock front formed by the impacting superwave, indicated by the arrows.*

center and the oncoming superwave onslaught.[25–27] Note that this bow shock is displaced westward from the main body of the remnant, as would be expected if the shock surrounds the remnant on the upwind side. The arrows indicate the direction from which the superwave cosmic rays would be approaching in this part of the Galaxy. Given its proximity to the superwave event horizon, it is not surprising to find that Cassiopeia A is *the brightest supernova remnant in the Galaxy at radio wavelengths.*

The Tycho remnant is also quite bright. It ranks as the fourth brightest Galactic supernova remnant at radio wavelengths. Observations of its radio emission spectrum show that its radiation is produced by electrons that match the observed galactic cosmic ray spectrum.[28] Rather than indicating that supernova explosions are responsible for generating the galactic cosmic ray background, as some have assumed, this data supports the view that these remnants are being externally energized by cosmic ray electrons that once originated in the core of our Galaxy.

Warning Markers. The Crab and Vela supernova remnants number among the small group whose explosions appear to have been triggered by the most recent major galactic superwave, the most intense superwave to pass through in the last 30,000 years. And recognizing the unique locations they have relative to us, we can only feel it is not just a coincidence that we see very intense pulsar flashes coming from them. The idea, then, strongly suggests itself that the Crab and Vela pulsars were placed in the heavens as *markers* to warn us about this past catastrophe. A precisely timed flashing signal is a universal archetype on our own planet for conveying a danger warning (e.g., yellow flashing lights for roadside construction hazards). It arouses attention much more effectively than a constant light source. Consequently, a pulsating beacon would be an ideal signal of choice if a galactic community wanted to convey a warning to novice civilizations of the existence of a galactic danger. The Vela pulsar currently flashes its warning 11.2 times per second, and as we approach the superwave event horizon to target the Crab pulsar, we receive a more urgent warning that pulses 29.8 times per second. This warning signal interpretation becomes even more credible when we realize that of all known pulsars the Crab and Vela pulsars are the brightest at optical, X-ray, and gamma ray wavelengths and among the top four brightest at radio wavelengths.

Let us now consider these various unique features that make the Vela and Crab pulsars the "king and queen" of the pulsar family.

The King and Queen of Pulsars

Bright Beacons at All Wavelengths. The Vela and Crab pulsars stand out from the crowd in several respects. First, they both produce very strong radiant energy emissions. At radio frequencies, the Vela pulsar is the brightest pulsar in the sky, outshining the others by several hundred times. The Crab pulsar is the most luminous pulsar in the sky, and also ranks as the fourth brightest when measured at radio frequencies—for example, 400 megahertz. In addition, both pulsars are unusual in that they emit pulses in the optical, X-ray, and gamma ray regions of the spectrum, where very few other pulsars emit detectable amounts of energy. Only three pulsars besides the Vela and Crab pulsars are known to emit optical pulses and one of those other three is the Millisecond Pulsar, the marker of the galactic one-radian point. Also, the Vela and Crab pulsars are among the group of just eight radio pulsars known to emit X-ray pulses and among seven known to emit gamma ray pulses. At gamma ray wavelengths, the Vela pulsar is the brightest source in the sky and the Crab pulsar is the fourth brightest gamma ray pulsar. When all these spectral regions are considered together, the Vela and Crab pulsars are found to be unique in that, among all known pulsars, they are the *only ones* that pulse in all of these spectral regions: radio, optical, X-ray, and gamma ray.

Interpulses. The Crab and Vela pulsars are also distinguished as being among the rare 1 percent of the pulsar population that produces interpulses, secondary pulses occurring midway between main pulses. The Vela pulsar produces interpulses only at higher energies, in the optical to gamma ray spectral range, whereas the Crab produces them also at radio wavelengths. The Millisecond Pulsar also happens to be one of the few pulse-interpulse beacons.

Giant Pulses. The Crab and Vela pulsars are also distinguished as being two of just ten pulsars that produce "giant pulses"—radio emission pulses that exceed the average pulse intensity by manyfold (figure 34). The Millisecond Pulsar and Eclipsing Binary Millisecond Pulsar in Sagitta are also among the few known to produce giant pulses,

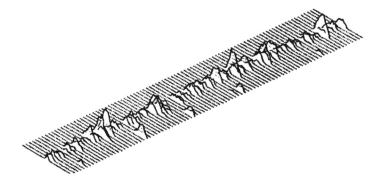

Figure 34. Stacked sequence of pulses representing 4 hours of observation of the Crab pulsar showing giant pulses. About 92 percent occur during the main pulse; the remaining 8 percent surface during the interpulse (courtesy of Gower and Argyle, Astrophysical Journal, *figure 1).*

although those produced by the Crab and Vela pulsars are far more intense. Giant pulses are observed only at radio wavelengths and in the Crab pulsar can rise up to 1600 times the normal pulse intensity.[29] Pulses ten times more intense than the mean recur at about 30 second intervals; those reaching 100 times the mean intensity recur about once an hour. Thousandfold giant pulses occur much more rarely.

In 2003, one group of radio astronomers announced the discovery that solitary subpulses forming the Crab pulsar's giant pulses occur in times as short as two nanoseconds.[30] Such a short duration implies that they are necessarily being emitted from a region less than two feet in diameter. This has led this research team to assert that their findings invalidate most previously proposed pulsar radio emission models. On the other hand, these results confirm the suggestion proposed in 2000 that pulsar signals are of artificial origin and are generated by particle-decelerating fields projected close to the surface of a stellar cosmic ray source. Although earlier I had suggested these might be 50 to 500 meters in diameter, in order to explain the Crab pulsar's giant pulses we would need to assume that these artificially created field disks are as small as half a meter in diameter.

When emitting a giant pulse, not only does the Crab pulsar become the brightest radio pulsar in the sky, exceeding even the intensity of the Vela pulsar, but its two-nanosecond giant subpulses have been proclaimed to be the brightest radio source in the universe. Were it not for its giant pulses, the Crab pulsar would be very difficult to detect at radio frequencies, as the radio emission background coming from its surrounding nebula is a hundred times brighter than the pulsar's average peak radio pulse intensity and so tends to mask the pulsar's signal. In fact, astronomers first discovered the Crab pulsar precisely because of these giant pulses.

The Vela pulsar also has interesting giant pulse features. It produces giant micropulses that typically last from 40 to 100 microseconds and occur just prior to the start of its main pulse and also broader micropulses, which occur more rarely, that last from 50 to 400 microseconds and appear on the trailing edge of its pulse profile.[31,32] This discovery marks the first time that micropulsation behavior has been reported for this pulsar. Vela, which normally is the brightest pulsar in the sky at radio wavelengths, can become up to 40 times brighter than its normal peak brightness during a giant pulse. Its giant pulses more closely resemble those coming from the Millisecond Pulsar in that they always occur at a defined pulse cycle phase, whereas in the Crab pulsar the giant pulses appear at varying phases anywhere within its pulse window.

Considering that the Crab, Vela, and Millisecond pulsars are among the few that produce giant pulses (<0.7%) and are three of just five pulsars that produce optical pulses, we are predisposed to forge a close associative link between them. The one-radian symbolism of the Millisecond Pulsar connotes an "arrow" flying away from the Galactic center and traveling a distance equal to the radial distance from the Galactic center to our solar system. The associative link established with the Crab and Vela pulsars encourages us to carry over this symbolic metaphor to envision the superwave's light-speed flight outward from the Galactic center, passing our solar system, close to that time detonating the Vela supernova, then continuing outward toward the galactic anticenter, where it subsequently detonated the Crab supernova and presently marks the position of its remnant shell. On the one hand we find the giant-pulse-emitting Millisecond Pulsar, through its strategic one-radian location, symbolizing the crest of the superwave's outward flight, and on the other hand we find the giant-pulse-emitting Crab pulsar currently at the forefront of the outward-traveling superwave event horizon.

The Crab pulsar also emits giant pulses during its interpulse phase, although these occur much less frequently than the main pulse giant pulses. One thing that has puzzled lighthouse model theorists is that the interpulse giant pulses and main pulse giant pulses do not occur together in the same pulse cycle, nor are the two correlated in any way. Further confusing the matter, giant pulses are not observed at optical, X-ray, or gamma ray wavelengths, even though the pulsar produces pulses at these high energies in synchrony with its radio pulses. Such "unnatural" behavior is the kind of thing that ET civilizations would try to engineer into their signals so that their communication beacons were not mistaken for natural sources.

Period Glitches. In addition to their intense output over a large portion of the electromagnetic spectrum, the Crab and Vela pulsars are unusual in that occasionally their precise, clocklike pulses can change dramatically. Pulsars normally have extremely constant pulsation periods, predictable to many significant figures, their periods slowly increasing over time at a very constant and predictable rate. However, the Crab and Vela pulsars are among the small percentage, 45 among the known pulsar population, in which this predictable rate of decrease occasionally changes in an abrupt manner. During such a "period

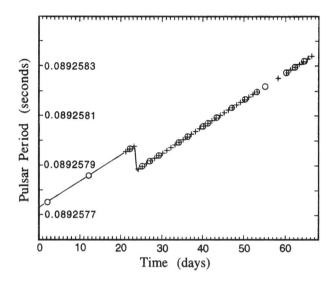

Figure 35. A 102-nanosecond-period glitch observed in the Vela pulsar on October 10, 1981 (after McCulloch, et al., Nature, *figure 1; by permission from* Nature, Macmillan Magazines Ltd.*).*

glitch," as it is called, the pulsation rate abruptly speeds up and afterward, over a period of several weeks, relaxes back close to its pre-glitch pulsation rate, whereupon the pulsar resumes its steady rate of period increase. Figure 35 shows one such period glitch for the Vela pulsar that caused Vela's period to shorten by about one part per million, a change of about 102 billionths of a second. This 1981 glitch was the fifth such event to occur since the pulsar's discovery in 1968. In all, the Vela pulsar experienced nine glitches during the 25 years following its discovery.[33] The Crab pulsar has undergone a similar number of glitches over a comparable period of time, but these have involved much smaller time shifts, of around a billionth of a second.

Depending on which pulsar is studied, there are varying ways in which a glitch and its subsequent period recovery will occur. Yet, the Vela and Crab pulsars happen to have remarkably similar glitch-recovery behaviors. Only one other glitching pulsar exhibits similar behavior, namely PSR 0525+21.[34] Curiously, this one happens to be the closest neighbor to the Crab pulsar, the two pulsars being separated from one another by just 340 light-years. Just a coincidence? We find that it lies about the same distance from Earth as the Crab pulsar, and in the plane of the sky is separated from the Crab pulsar by just 1.3 degrees of arc (see fig. 36). Only one other pulsar is found within 5 degrees of the Crab pulsar, but it lies several thousand light-years farther away behind the Crab Nebula. The probability of two pulsars being located so close together in space and happening to exhibit the very rare period glitch phenomenon is very small, less than one chance in 44,000. The chance is even smaller that both would also exhibit similar glitch-recovery behaviors, found in just two other glitching pulsars, and that both would be positioned so close to the galactic anticenter. This would calculate to one chance in a billion. If we take into account the probability that one of these two, the Crab pulsar, would be such a highly unusual pulsar, one displaying the rare characteristics of producing giant, optical, X-ray, and gamma ray pulses, we arrive at the extremely improbable value of less than one chance in 3×10^{18}.

Period glitches are among the many pulsar signal-ordering characteristics that have confounded astronomers. Working in terms of the troublesome assumption that a pulsar's pulses are produced by the "lighthouse flashes" from a rotating neutron star, theorists have

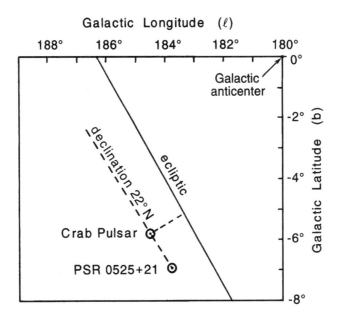

Figure 36. A galactic coordinate map showing the positions of the Crab pulsar and PSR 0525+21. A line projected through them is within 0.01 degree of being exactly parallel to the celestial equator and within a few degrees of paralleling the ecliptic. Their positions coincide with declination 22° north.

been forced to conclude that a speed-up glitch is produced by a sudden increase in the neutron star's spin rate. They suggest that this "spin-up" might be due to a sudden shrinkage in the neutron star's radius, causing, in turn, an abrupt gain in its angular momentum and spin velocity. But how could such a "star quake" occur and not have any major effect on the star's long-term spin-down rate? As seen in figure 35, shortly after the Vela pulsar's glitch, the pulsar's period continued to increase at nearly the same rate as before the glitch occurred. How could a pulsar's pulse period be so precisely timed if at any moment this period is capable of undergoing such a drastic change? Current neutron star models are far from providing an adequate explanation. Moreover, PSR 0525+21 has presented a particular challenge to lighthouse model theorists, as astronomers have concluded that its glitching cannot be due to star quakes and that some other explanation must be sought.

Occasional period glitches make a lot of sense if pulsars are instead extraterrestrial communication beacons. Unusual changes of this sort could be intentionally incorporated both to attract our attention to the pulsar and to confound any attempts we might make to devise a natural explanation. When a repeating signal of such high precision suddenly shifts its period, even by a very small amount, and immediately afterward resumes its previous precise rate of period increase, this is something that certainly will not be missed by observant astronomers. One other curious discovery: astronomers have found that the Earth is being showered by faint pulses of 100-trillion-electron-volt cosmic rays coming from this part of the sky and having a period close to that of PSR 0525+21.[35] Is this cosmic ray message a warning to us of things to come?

By itself, the Crab pulsar is unique and attention-getting. But seeing another pulsar with such similar glitching properties and positioned so close to it, our interest becomes heightened even more. As we will see shortly, these two pulsars make special alignments both to the Earth's orbital plane and to its celestial equator. This is one of many "coincidences" relating to the Crab Nebula and Crab pulsar that we will explore in the next chapter.

Warning of an Impending Superwave?

In the fall of 2003 radio astronomer Scott Hyman of Sweet Briar College discovered a very unusual transient radio source which has become designated as GCRT J1745–3009.[35] The source lies just 1.1

degrees southwest of the Galactic center at galactic coordinate position ℓ = 358.891±0.001, b = −0.542±0.001. It lies about as far away as the Galactic center so it is physically close to the Galaxy's core. Hyman reports recording five radio bursts from the source over the approximately six-hour viewing period. These were found to recur approximately every 77 minutes, each burst lasting for 10 minutes. The pulses, which were observed at a radio frequency of 330 MHz had an intensity of slightly more than 1500 milliJanskys, making them the second brightest pulsed radio beacon in the sky. But when astronomers attempted to re-acquire its signal, they discovered it had vanished. A search of earlier data records turned up only one other detection where a single pulse had been recorded. Consequently, it is considered to be a transient source that spends most of its time in an off state.

The 77 minute pulse period for this source is far longer than the longest period observed for any radio pulsar, which is 11.7 seconds. But the unusual regularity of the pulse cycle period makes it reminiscent of a pulsar. Such regularity has not been seen in other transient radio sources. Even more curious, there was no evidence that the emission was accompanied by X-ray or gamma-ray emission, as is seen in other burst sources. It is also unique in that its broadband radio emission is coherent, resembling that coming from a radio pulsar or free-electron laser.

Needless to say GCRT J1745–3009 is an extremely unique object. Like the Millisecond Pulsar, it stands out as a one of a kind. Due to the relatively scant data set, there is no way to say at this point if its signal might contain the highly complex ordering seen in a pulsar. However, if it is to be placed in the same class as pulsars and considered as a beacon of artificial origin, of all such beacons it would be the one that is positioned closest to the Galactic center. Is this just a highly unusual radio source that happens to lie close to the Galactic center, or is this an intentional warning marker set up to call attention to the superwave message of the pulsars?

SIX

SKY MAPS OF A CELESTIAL DISASTER

A Star Chart of the Sagitta Constellation?

Each pulsar has a unique pulsation period. This can be nearly as small as one and a half milliseconds or as long as 11.7 seconds, with most being around several tenths of a second in length. In addition, each pulsar has a unique rate at which its characteristic period, P, slowly changes over time, a quantity that is termed the *period derivative,* symbolically represented as dP/dt or alternatively as $\dot{P}$. Period derivatives can range from as small as 0.05 picosecond per year ($\dot{P}= 1.5 \times 10^{-21}$ seconds per second) to as large as about 10 milliseconds per year ($\dot{P} = 4.2 \times 10^{-10}$ seconds per second). Yet despite these wide-ranging values, a pulsar's period changes at an exceedingly constant rate, period derivatives having precisions that reach as high as eight significant figures. The pulsation rates of most pulsars are gradually slowing down, but the pulsation rates of a small subset, 24 of the known 1,533 pulsars, are gradually speeding up. To see how the various pulsation timing characteristics compare with one another, astronomers sometimes graph a pulsar's period against its period derivative on a logarithmic plot such as that shown in figure 37.

The three pulsar coordinates marked by boxes in the upper-left-hand corner of the graph plot the logarithmic period and period-derivative coordinates (log P–log $\dot{P}$) for the Crab pulsar, Vela pulsar, and Vulpecula pulsar (PSR 1930+22). All three are associated with

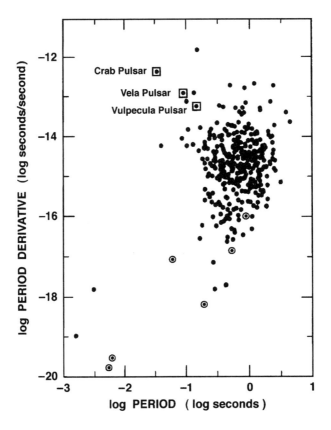

Figure 37. Pulsar period P (horizontal axis) plotted against pulsar period derivative Ṗ (vertical axis) on a logarithmic graph for 305 pulsars (after Dewey, et al., Astrophysical Journal, *figure 2b). The squares mark the coordinates for the Crab, Vela, and Vulpecula pulsars.*

supernova remnants, and all three are relatively fast pulsars with periods of 33.4, 89.3, and 144.5 milliseconds, respectively. Interestingly, the layout of the coordinates for these three pulsars calls to mind the appearance of the Sagitta star constellation. Identify galactic longitude, ℓ, with log P, where log P is instead plotted to increase from right to left in figure 37, and then compare figure 37 with figure 15, in which galactic longitude similarly increases from right to left. Also identify galactic latitude with log Ṗ where b becomes increasingly negative with decreasing log Ṗ (top to bottom in figure 37). The relative spacing of the three P–Ṗ points and their alignment relative to the vertical Ṗ axis is then seen to be very similar to the spacing of Delta,

Gamma, and Eta Sagittae and the alignment of these stars with respect to the galactic equator. Delta and Gamma Sagittae are the stars that form the shaft of the Celestial Arrow, and Eta Sagittae is the target star toward which Sagitta is flying. By identifying the P–$\dot{P}$ coordinates for the Crab and Vela pulsars with the present galactic longitude and latitude (ℓ, b) coordinates of Delta Sagittae and Gamma Sagittae, it is possible to derive a mathematical relation for converting P–$\dot{P}$ coordinates into galactic coordinates; see text box.

The Vulpecula pulsar (PSR 1930+22) does not have anywhere near as many distinguishing features as the Crab and Vela pulsars. It is only of average radio brightness; it does not emit pulses in the optical, X-ray, or gamma region of the spectrum; and its period does not exhibit glitches. Nevertheless, like the Crab pulsar, it is superimposed on a supernova remnant shell and is located near the 12,150-B.C.E. superwave event horizon, the ellipsoidal boundary that demarks the observed location of the superwave cosmic ray volley that passed the Earth 14,150 years ago (figure 31).[1,2] However, since the Vulpecula pulsar's background remnant is at least 300,000 years old, the supernova explosion that produced it could not have been triggered by this particular superwave.

The Vulpecula pulsar is also distinctive in that it is physically located just 0.13 degree from the Galactic center's one-radian longitude meridian, closer than any other pulsar. In this respect, it is very fitting that this pulsar should represent Eta Sagittae (η Sge) on the P–$\dot{P}$ map, since the Sagitta constellation pictures this as the star toward which the arrow is flying. When the entire pulsar population is plotted, additional pulsar coordinate points crowd in near the Vela and Vulpecula pulsar P–$\dot{P}$ points, somewhat confusing the clarity of the "Sagitta map." However, the Crab, Vela, and Vulpecula pulsars are sufficiently distinctive to stand out from the rest.

An Event Chronometer

As explained in chapter 4, the constellation of Sagitta (the Celestial Arrow) is a key part of the zodiac star lore cipher in that it uses the radian concept to convey the idea that a cosmic ray outburst, or superwave, traveled all the way from the Galactic center to our solar system, arriving on the date indicated by Sagittarius's arrow indicator.

Converting P–Ṗ Coordinates into Galactic Coordinates

Using the P and $\dot{P}$ values that these pulsars had on May 1, 1992 (table 3), we obtain the following relation:*

$$\ell° = 63.3687 + 5.1495 \times \text{LOG } P$$
$$b° = 39.6429 + 3.4766 \times \text{LOG } \dot{P}$$

Now, by taking the P and $\dot{P}$ values for PSR 1930+22 from table 3 and plugging them in to the above equations, we can see how closely this "pulsar map" predicts the galactic longitude and latitude coordinates for the constellation star Eta Sagittae. Doing so, we get a map coordinate position of $\ell = 59.04°$, $b = -6.39°$, which deviates by only 0.2 degree from Eta Sagittae's actual sky position of 59.18°, −6.23°. If we instead construct our pulsar map using the P and $\dot{P}$ values the Crab and Vela pulsars had 23 years earlier, in 1969, we find that the pulsar map predicts the same map coordinate position for Eta Sagittae even though the P and $\dot{P}$ values for the Crab and Vela pulsars have each changed considerably in the intervening time.

TABLE 3. PERIOD AND PERIOD DERIVATIVES FOR THREE UNIQUE PULSARS

Pulsar	Period(s)	Period Derivative (s/s)	Date
Crab	0.033403347	4.209599×10^{-13}	1 May 92
Vela	0.089298530	$1.258 \pm 0.008 \times 10^{-13}$	1 May 92
Vulpecula	0.144457105	5.75318×10^{-14}	1 May 92

*The values of the constants used to convert P and $\dot{P}$ into ℓ and b depend on the time unit used to measure the pulsar periods and on the base of the logarithm used. To produce the pulsar map, it is not important what measurement system we use provided that we use it consistently. In our calculations we adopt seconds for a time measure and base 10 for our logarithm operation, in accordance with figure 37. Also, the conversion constants are based on the P and $\dot{P}$ values the Crab and Vela pulsars had on Julian date 2,448,743 (May 1, 1992).

Knowing this, we are led to consider whether the pulsar map representation of Sagitta may be conveying a message about this recent galactic superwave and whether it too may be an *event chronometer* similar to Sagittarius's arrow. Recall Sagittarius's arrow not only points out the location of the Galactic center but also serves as a chronometer that indicates the date when the most recent major superwave began passing the Earth (see chapter 4). That is, the Archer's aim gradually departs from his designated target, the Heart of the Scorpion, because the sky positions of the nearby stars forming the Archer's arrow (γ Sag and δ Sag) slowly change as these stars drift relative to distant background stars. However, when we regress the positions of these stars back in time, we find that the arrow would have been on target around 16,000 years ago, which gives the date when a Galactic center outburst would have first become apparent to Earth observers.[3]

The Vela pulsar coordinate point, which represents the head of the pulsar map "arrow," is drifting to the right in the $P-\dot{P}$ map due to the gradual lengthening of Vela's period at the rate of 1.25×10^{-13} seconds per second. Thus, it is natural to instead look to the past and ask: *On what past date would the Vela pulsar have had a pulsation period equal to the present period of the Crab pulsar?* Knowing that the Vela pulsar is currently pulsing about 2.67 times slower than the Crab pulsar (11.2 pulses per second as compared with 29.8 pulses per second), and knowing the rate at which the Vela pulsar is slowing down, through a simple calculation we find that Vela would have had the Crab pulsar's present pulsation rate 14,100±100 years ago.* Interestingly, the superwave that triggered the Vela and Crab Nebula supernova explosions would have been passing Earth on this designated date.

As noted earlier, polar ice-core data indicates that galactic cosmic ray intensity outside the solar system had reached a peak 14,150±150 years ago and had begun its rise to that climax some centuries earlier, perhaps 14,300±200 years ago. Noting that this peak was one of the most prominent of the last ice age, the possibility suggests itself that the Vela and Crab pulsars are highly accurate chronometers of ETI origin

---

*This time-span calculation presumes that the Vela pulsar's rate of period change $(\dot{P})$ does not change over time. Although its value does decrease with time, it rapidly resets to a higher value when a pulsar glitch occurs. Due to the irregularity of this glitch-reset phenomenon, more data must be collected before we can determine whether the average $\dot{P}$ in fact changes over the long term.

that have been placed in space to inform us about this key superwave date.

The geologic record registers a number of other things happening around the time of this superwave passage.[4] The Earth's climate began to warm up very rapidly beginning about 12,700 B.C.E. and reached a temperature peak around this 12,150-year-B.C.E. date, this period of unusual ice-age warmth being known as the Bölling Interstadial. This warming could have been initiated by cosmic dust and gas the super-wave brought into the solar system, which in turn increased the Sun's activity and also increased the amount of solar radiation the Earth received through scattered light. Other data indicate that at this time the glacial ice sheets were melting at their fastest rate and causing wide-spread continental flooding. This date also marked the beginning of a major episode of mammal extinction that climaxed 1500 years later. As noted in chapter 4, this was most likely the result of a solar flare cosmic ray disaster. In addition, around 12,200 B.C.E., the Earth's north magnetic pole abruptly flipped southward for a period of several decades, possibly caused by the impact of intense solar flare outbursts from an excessively energized Sun.

It is quite fitting that the Crab, Vela, and Vulpecula pulsars, which mark sites where major celestial explosions once took place, would be used to transmit a picture of Sagitta, a constellation that in our star lore symbolizes the passage of a catastrophic core explosion superwave.[5] Moreover, to the extent that a supernova causes the death of a star, pulsars associated with them might be viewed as warning beacons. It is also appropriate that ETI civilizations would utilize communication beacons that use cosmic ray beams to emit synchrotron radiation, the same type of radiation that is emitted from superwave cosmic rays.

A Celestial Memorial
to a Terrestrial Cataclysm

Probably one of the most perplexing discoveries is that the present dimensions of the Crab supernova remnant are proportional to those of the 12,150-B.C.E. superwave event horizon as it would have been viewed by us 950 years ago, at the time of the Crab Nebula supernova explosion. Thus, the Crab Nebula appears to be an immense three-dimensional celestial map that accurately portrays the ellipsoidal con-tour of the superwave wavefront as seen around this time. Compare the shape of the nebula shown in figure 30 with the oval event horizon shown in figure 31. Like the superwave event horizon, the Crab Nebula has the shape of a prolate ellipsoid, one that is elongated like an American football rather than squashed like a discus. In addition, the nebula's major and minor axes happen to have about the same length ratio as the major and minor axes of the superwave event horizon ellipsoid. Measuring about 13.8 by 10.6 light-years in extent, the Crab Nebula has an axial ratio of 1.30. By comparison the 12,150-B.C.E. superwave event horizon currently has an axial ratio of 1.27. But in 1054 C.E. this event horizon would have had an axial ratio of 1.295, very close to the Crab Nebula's 1.30 ratio.* Thus, the Crab Nebula approximates a 1:4500 scale model of the shape the superwave event horizon would have had in 1054 C.E.

Furthermore, the orientation of the nebula's major axis is such that it is parallel to the galactic plane and *perpendicular* to the Galactic center–anticenter axis, which extends from $\ell = 0°$ to $\ell = 180°$. On the other hand, the long axis of the superwave event horizon (as seen from Earth) is instead oriented in line with the Galactic center–anticenter axis. In other words, this ellipsoid "scale model map" is rotated so that its major axis is oriented almost exactly perpendicular to the major axis of the superwave event horizon. Just a coincidence? If its long axis was not positioned in this manner, the Crab Remnant would appear to us as a circle, making its ellipsoidal shape difficult to discern. Since the long axis of the Crab Nebula is aligned perpendicular to our viewing direction, the nebula's unique elliptical shape is made *most apparent from our observing location.*

Because the Crab remnant consists of material ejected from a central explosion, it serves as an appropriate symbolic representation of the galactic superwave whose cosmic rays were similarly explosively ejected. Moreover, the cosmic ray electrons that are magnetically trapped within the Crab Nebula's ellipsoidal shell radiate a synchrotron radiation continuum much like those journeying outward in the ellipsoidal shell that forms the superwave event horizon. From our location

---

*The major-to-minor axis ratio of this event horizon progressively decreases with the increase in time that has elapsed since its Earth passage date. More specifically, this ratio is given as $(23 + t)/(46t + t^2)^{1/2}$, where t is the time in thousands of years since the date of superwave passage. This assumes that the Galactic center lies 23,000 light-years away (~7.1 kiloparsecs). If the Galactic center distance is instead figured to be as great as 26,000 light-years, this ratio would calculate to be 1.34 in 1054 C.E. and 1.31 at present.

The Crab Nebula: Scale Model
of the Superwave Event Horizon

The Crab supernova remnant shell serves as a scale model of the 12,150-B.C.E. superwave event horizon for the following reasons:

- It has a prolate spheroid shape, like the event horizon.
- It has the same 1.30 axial ratio the event horizon would have had on the date of the Crab supernova.
- The nebula's minor axis is orientated perpendicular to the galactic plane, as is the event horizon minor axis. The nebula's major axis is oriented parallel to the galactic plane, as is the event horizon major axis, but is rotated by ninety degrees to the event horizon major axis, allowing it to be seen broadside to our viewing direction.
- Like the superwave event horizon, the Crab remnant emits synchrotron radiation from superwave cosmic rays traveling outward from the Galactic center.

within the superwave shell, we observe this superwave radio emission as a diffuse background radiation, which astronomers have termed the galactic radio background radiation.

In addition, as we have already mentioned, the Crab, Vela, and Millisecond pulsars share certain similarities that occur very rarely among pulsars, and which encourage us to conceptually interrelate these three pulsars. The one-radian symbolism of the Millisecond Pulsar, combined with the Crab and Vela pulsar $P–\dot{P}$ mapping of the Sagitta arrow constellation with its 12,100-year-B.C.E. date, evokes the idea of the superwave event horizon's radial light-speed flight away from the Galactic center, a phenomenon that is further corroborated by the dates for the Crab and Vela supernova explosions. All of this symbolism together reinforces the idea that the Crab Nebula forms a map of the 12,150-B.C.E. superwave event horizon.

Could these similarities of the shape and orientation of the Crab Nebula to that of the superwave event horizon just be a series of unusual happenstances of nature? Or, do they perhaps indicate that the Crab supernova was intentionally engineered, a cosmic fireworks display performed for the benefit of our civilization? The second possibility boggles the imagination. To induce a stellar supernova and to then shape the resulting explosion to conform to the precise dimensions of the ellusive superwave event horizon implies that these cosmic artisans have mastered the ability to manipulate nature's energy processes on an inconceivably grand scale. This calls to mind the vast astro-engineering feat that was necessary to place and orient the Eclipsing Binary Millisecond Pulsar at its unique sky position.

In addition, the Crab supernova occurred in a unique location relative to the solar system. As mentioned before, of all young supernova remnants (those less than 2,000 years old), the Crab Nebula is the closest one to our solar neighborhood and is also the closest to the galactic *anticenter* (fig. 28). Recall that the galactic anticenter is the direction in the sky lying directly opposite to the Galactic center. Hence, the anticenter direction relative to background stars will depend on the viewer's location in the galactic disk. When considered in view of the Crab Nebula's unusual brightness, its unique ellipsoidal shape, and its careful orientation parallel to the galactic plane and perpendicular to our line of sight, it becomes difficult to imagine that all this was purely accidental.

Also, the Crab Nebula is carefully placed relative to key constellation stars. It is located in the Taurus constellation very close to the line that extends between the two stars marking the tips of Taurus's horns, and lies slightly more than one degree from Taurus's southern horn tip (fig. 38). As I have shown in *Earth Under Fire*, the charging Celestial Bull in ancient Egyptian myth and Minoan custom signified the galactic catastrophe that long ago brought havoc to the Earth.[6] Taurus, who is shown charging toward the galactic anticenter, symbolizes the superwave as it advances away from the Galactic center and the tips of his horns signify the superwave's forefront. Orion, who confronts the Bull with raised shield and points out the galactic anticenter with his club, is by tradition said to be a memorial set in the sky to honor those who died in this global catastrophe. It is very interesting that the Crab Nebula is so appropriately placed in this symbolic pictorial reenactment of humanity's encounter with the galactic superwave. Just as the nebula is situated in the sky at the forefront of the charging Bull, so too it is presently situated at the forefront of the superwave and is being impacted by outwardly traveling galactic cosmic rays. Did an advanced extraterrestrial civilization visit our planet in ancient times and teach our ancestors about the constellation lore of Taurus and Orion (Osiris), knowing that this carefully engineered supernova, which took place in 5530±30 B.C.E. would one day 6,585 years in the future become visible on Earth?

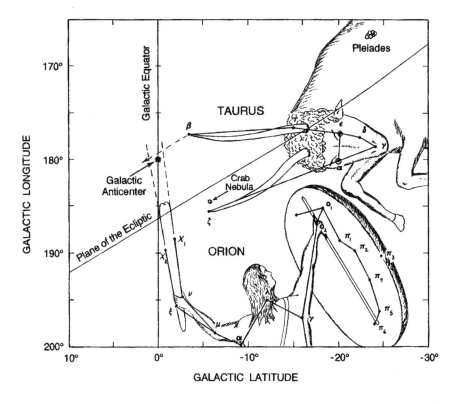

Figure 38. A map showing the constellations of Taurus and Orion with the position of the Crab Nebula marked out near the tip of the Taurus's lower horn (LaViolette, Earth Under Fire, *figure 3.3).*

Furthermore, the Crab Nebula is distinguished as being the only galactic supernova remnant outside of the galactic nucleus to lie so close to the Earth's ecliptic plane. Another coincidence? The Crab pulsar is positioned just 1.29 degrees below the ecliptic. Its close proximity to pulsar PSR 0525+21 also draws attention to the nebula's ecliptic position. As noted in the previous chapter, these two pulsars not only are unusually close to one another, but they also share the rare phenomenon of pulse glitching and a type of glitching seen only in two other pulsars. As it would be extremely unusual to find two such glitching pulsars so close together, less than one chance in a billion, it is natural to examine their positional relation relative to each other.

What we find is quite remarkable. A line connecting these two pulsars comes within a few degrees of being parallel to Earth's ecliptic plane (fig. 36) and currently *deviates by 0.01 degree from being exactly parallel to the 2000* C.E. *celestial equator* (it will be exact in 2100 C.E.). Moreover, the two pulsars are presently separated from one another by 1.31° of arc, which is within 0.02° of equaling the Crab pulsar's present 1.29° latitudinal deviation from the ecliptic plane. Could PSR 0525+21 have been purposely positioned near the Crab pulsar to call attention to the Crab supernova remnant's position relative to our ecliptic plane?

These unusual positional arrangements compel us to consider whether the Crab Nebula supernova explosion is purposely being marked for our viewing benefit and not merely for civilizations in our general galactic locale. If so, is humanity considered so important that this galactic community would go to these grand efforts to produce this immense nebula map specifically for us?

The location of the Crab Nebula may be personalizing its message for us in yet another way. The distance from the Earth to the Crab Nebula, which is measured to be 6,585±30 light-years, also designates a light travel time period that amounts to about one fourth of the time taken for the Earth's precessing poles to complete one circuit of the ecliptic, a full polar "Great Cycle" taking approximately 26,000 years. Interestingly, the time required for the 12,130-B.C.E. superwave event horizon to travel the Galactic center radial distance from our solar system to the Crab Nebula's progenitor star, a total of 6,520 years, when added to 6,585 years, the time required for light from the Crab supernova explosion to travel the intervening distance back to us, yields a round-trip travel time of 13,105±60 years. By comparison, in 12,130 B.C.E., when that supernova-triggering event horizon was passing the solar system, the time required for the Earth's pole to precess 180 degrees, half of a Great Cycle period, averaged 13,070 years.

If this explosion was purposefully arranged, it suggests that its perpetrators *knew not only the plane of Earth's orbit around the Sun, but also the rate at which the Earth's poles precess.* Taken out of context, the idea of an intelligently sited supernova would sound absurd. But viewed in terms of the other "coincidences" noted earlier, it may be a thought worth considering.

Let us further investigate the Crab Nebula's position along the ecliptic from the standpoint of whether it might represent an intentional time marker. That is, the Earth's poles precess along the ecliptic at a very regular rate, and this gradual polar movement causes the seasons to change relative to the background constellations. By noting the position of the moving vernal equinox relative to stationary markers such as the boundaries between zodiac constellations, ancient cultures were able to designate the passage of zodiacal ages and the dates of significant events that took place thousands of years past. In a similar fashion, a galactic community may have intentionally detonated the Crab supernova to mark a particular date on the Great Cycle chronometer.

As shown above, the Crab Nebula's singular location appears to implicate a kind of purposeful arrangement, as though these sky artisans carefully selected the location of the supernova progenitor star in order to convey to us that they knew the location of our ecliptic plane and the rate at which our planet's poles have been precessing. If so, could they also have chosen this particular location because its longitudinal position along the ecliptic might call our attention to a particular date in the past? If we regard the vernal equinox as the Great Cycle time indicator, as is traditional practice, we find that it would have been positioned at the Crab Nebula's ecliptic longitude (84.1°) in the year 4120±25 B.C.E. This date is not very significant from an astronomical or geological standpoint other than being 1,400 years after the actual date on which the Crab supernova occurred 6,520 light-years from us. Although, interestingly, this date does fall within 120 years of 4240 B.C.E., the zero date for the ancient Egyptian Sothic calendar.*

But let us instead consider the date that comes one quarter of a precessional cycle earlier, a time when the winter solstice coincided with the Crab Nebula marker and the vernal equinox fell between the constellations of Virgo and Leo. We now arrive at a more meaningful date

---

*The Egyptians had two calendars. Their Sothic calendar, which was based on the heliacal rising of the star Sirius, accurately measured out the year as 365 ¼ days. Their "vague year" calendar measured out the year instead as a whole number of 365 days. As a result of this discrepancy, the two calendars would begin on the same day only once every 1,460 years. On that day of calendric coincidence, the Egyptians would celebrate their New Year at their sacred Temple of Hathor at Dendereh, the first such New Year being on 4240 B.C.E., the beginning of the Sothic calendar (Schwaller de Lubicz, *Sacred Science* [Rochester, Vt.: Inner Traditions, 1982], 174). Interestingly, the zodiac found on the ceiling of this temple displays among its constellations a New Year date marker figured as a cow with a star between its horns. The cow symbolized Hathor, the cow-headed goddess, a form of Isis, and the star presumably denoted her star Sothis (Sirius). Was it foresight that caused the ancient Egyptian priests to depict a cow with a star between its horns as their New Year marker, the first New Year falling near the date indicated by the Crab supernova ecliptic marker, a supernova that occurred between the horns of Taurus? Or did visitors from above tell them that the significance of this star would be understood one day three millennia in the future?

of 10,740±50 B.C.E. This correlates with the climax of the terminal Pleistocene extinction (10,750±100 B.C.E.), the most significant large-animal mass extinction episode to occur since the demise of the dinosaurs. The extinction was not due to overzealous paleolithic hunters as some have theorized. The event undoubtedly elicited an enormous human toll as well. Both geologic evidence and legend indicate that, at the time of this catastrophe, the Earth endured a period of extreme heat that caused vast deluges of glacial meltwater to sweep over the continents. One theory suggests this conflagration occurred when the Earth became engulfed by a hazardous solar coronal mass ejection.[7] The associated barrage of solar flare cosmic ray particles would account for the instability of the Earth's magnetic field during this period as well as for the high levels of atmospheric radiocarbon and fifty-times-higher solar cosmic ray intensity recorded in moon rocks. The Sun appears to have been aggravated into this active state by cosmic dust and gas injected into the solar system by a passing superwave, the same superwave currently energizing the Crab Nebula.

At the time of this mass extinction, the vernal equinox was positioned at the boundary between the constellations of Virgo and Leo, designated by the boundary stars Omega Virginis and Beta Leonis. Our ancestors attached such great importance to this episode that they memorialized its time of occurrence by using star constellations to reference this vernal equinox position. For example, the ancient Egyptians and ancient Greeks both chose this ecliptic boundary as the starting point for their precessional Great Cycle calendar. Also, one ancient Greek flood myth records this date by describing how Zeus punished mankind with a deluge immediately after the goddess of harvest, Astraea, departed from Earth. Because she was represented by the Virgo constellation, her departure refers to the time when the Age of Virgo ended and the vernal equinox was about to pass into Leo.[8] Remembrance of this Virgo–Leo boundary date has also been preserved in the floor mosaic found in the sixth-century-A.D. Beth Alpha Synagogue. It depicts a zodiac with the spring equinox positioned at the Virgo–Leo boundary. The center of the mosaic displays the sun god Helios with his four steeds, calling to mind the ancient Greek conflagration myth about Phaethon and the sun chariot.

The Crab Nebula, then, through its unique sky position near the tip of Taurus's southern horn, may be memorializing one of the worst mass extinction episodes to occur in recent geologic history and which, according to legend, involved the loss of many human lives. But the Crab Nebula also contains a unique pulsar that is part of the phenomenal interstellar superwave message we have been analyzing. Thus, if pulsars have been created by a network of galactic civilizations, it seems these civilizations were well acquainted with the solar flare cataclysm that affected our planet at the time of the last superwave passage.

A Superwave Shield?

One of the most puzzling features of the Crab supernova remnant is the tubular jet of ionized gas that is seen jutting outward from its northern side (fig. 39). This immense tube extends about two and a half light-years beyond the perimeter of the nebula and has a diameter measuring roughly one and a half light-years. Its unusually straight and parallel sides contrast remarkably with the irregularly twisted filaments that make up the rest of the nebula.

In their 1982 *Astrophysical Journal* article reporting on this "mysterious jet," the astronomers Theodore Gull and Robert Fesen raised the following perplexing questions:

> First, why is the jet the only filamentary extension outside an otherwise well-bounded nebula, and second, why does it have such an organized structure in a largely chaotic and amorphous-appearing supernova remnant? Because of its appearance, it is not clear that the jet has undergone the same physical processes which shaped the rest of the nebula. Though its structure suggests an energetic origin, it is not obvious that the jet is directly related to the A.D. 1054 supernova event since it is not in radial alignment with either the center of expansion or the pulsar . . . clearly more information regarding the jet's physical properties . . . is required before one can realistically attempt to understand the jet's structure and origin.[9]

Moreover, observations of material in the jet's northern tip show it to be receding from us at a relatively low velocity of 100 to 150 kilometers per second, which contrasts with the high velocity of the nebula's outer envelope, which expands at rates ranging from 1500 to 2200 kilometers per second. As a result, astronomers have concluded that the jet must be oriented nearly in the plane of the sky, with most of its material moving perpendicular to our line of sight.[10] But long-term

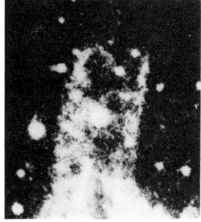

Figure 39. The Crab Nebula and its optical jet imaged with special green and red filters that pass light emitted from doubly ionized oxygen as well as from ionized hydrogen and nitrogen. Right image shows a magnified view. (Gull and Fesen, Astrophysical Journal, *figures 2b, 2c).*

observation of 14 luminous filaments in the jet shows that this material is moving directly away from the nebula's center of expansion, like the filaments in other parts of the nebula, and not in the direction of the jet's long axis. This has puzzled astronomers who were expecting to find a reasonable way to explain the jet, as their observations indicate it cannot have been formed by the nebula's explosion. The astronomers Robert Fesen and Bryan Staker comment about this in their 1993 *Monthly Notices* paper:

> [T]he expansion directions of the 14 measured jet features appear most consistent with a simple radial expansion directed away from the centre of the remnant . . . It seems, therefore, that all image data available to us point to a radial-like expansion, in direct contradiction to the expected non-radial motions which the optical structure of the jet seems to demand . . . If correct, it means that the parallel-edged jet structure is not the result of peculiar non-radial kinematics along the northern boundary of the Crab.[11]

As we can see, astronomers are thoroughly perplexed as to why the jet is not aligned with the nebula's expansion center, for if it was, its formation might be explained by the natural process of nebular expansion. Nevertheless, such off-center alignment is precisely the kind of thing an

ETI civilization would do to ensure that the recipients of their message did not mistake it for a natural phenomenon, and as we see, the Crab Nebula jet certainly did get the attention of our astronomers.

However, if we look carefully at the orientation of the jet, which makes approximately a 40° angle to the nebula's major axis, we see that it may be purposefully positioned. Its eastern edge (left side in the photo) *is aimed directly at the southern focus of the nebula's ellipsoid* (see fig. 40). There is no natural explanation for such an alignment; explosion ejecta would always scatter outward from the explosion center point, not from one of the foci of the ellipsoid that the explosion happened to form.

It is not clear whether this siting is relevant in the context of the scale model Crab Nebula map discussed earlier. Recall that the two foci of the ellipsoid would represent respectively the Galactic center and the Earth. Thus, if the southern ellipsoid focus is interpreted as representing the Earth, the map would be showing the jet as symbolically aiming toward the Earth.

If the Crab supernova explosion was purposely engineered by a highly advanced extraterrestrial civilization, this type of unusual organized feature does not come as such a surprise. But the question that naturally arises is How did they form this? Could the jet be an example of a cylindrical force field shield deployed to intercept super-wave cosmic rays? Radio telescope studies show that the jet indeed emits polarized synchrotron radiation, which indicates that it is being

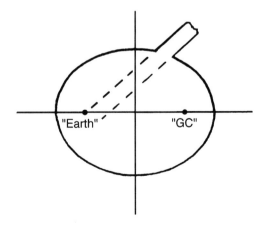

Figure 40. Sketch showing the location of the foci of the Crab Nebula's ellipsoid and the relative orientation of its optical jet.

energized by cosmic rays that it is trapping in its magnetic field (see radio jet in fig. 30).[12] Could the civilization that engineered the Crab supernova remnant and its unique pulsar beacon also be demonstrating for our benefit a technology that might one day protect us from the onslaught of the next superwave? Note that the Crab Nebula map shows this jet or "shield" symbolically intervening between the Earth and the Galactic center. Perhaps these sky artisans are trying to tell us we should be prepared to produce such a shield to shelter ourselves from the Galactic center.

This same shield metaphor has been left for us in our constellation lore. As seen in figure 38, Orion holds up a shield to ward off the Bull of Heaven (the galactic superwave), which arrives from the Galactic center and charges toward the galactic anticenter. We find the shield symbol also appearing in the constellation of Centaurus. The one-radian metaphor conveyed by both the Sagitta and Crucis constellations, which are positioned on opposite sides of the Galactic center, leads us to symbolically "fold" these two one-radian vectors toward us about the Galactic center "fulcrum point" so that these constellational one-radian vectors coincide at our immediate locale to form a vector extending from the Galactic center directly toward our solar system. We may then picture Centaurus immediately in front of our solar system facing the Galactic center with his shield upraised, sheltering us from the onslaught of galactic superwave cosmic rays represented by the outward flying arrow Sagitta.

Are they trying to tell us that in time of need they are willing to come to our aid during a superwave cataclysm and, if asked, help us by energizing a similar shield around the solar system? Or are they bound by the "Prime Directive" to avoid overt interference in human affairs on matters relating to survival? Is the galactic superwave event considered to be a test of a civilization's social advancement? If a planetary civilization were to successfully defend itself against such a galactic disaster and survive as a peaceful society without lapsing into a chaotic "dark age" period, would it then be considered worthy to be removed from quarantine and admitted to the galactic federation? Does it want only to warn us of the existence of the superwave phenomenon and to inform us that a technology exists which we can use to defend ourselves? In chapter 8 we will investigate the feasibility of using cutting-edge force field beam technology now under development to deploy a space shield similar to that being demonstrated for us in the vicinity of the Crab Nebula.

Cosmic Synchronicity?

For members of the Galactic ETI community to make such a major effort to create the Crab Nebula celestial map for our viewing, they must have known that our planet was inhabited. Let us suppose these artisans resided not far from the Crab Nebula and learned about our planet by exchanging faster-than-light messages with other members of their galactic "club" who lived in our solar neighborhood. Or perhaps they had the capability of superluminal space travel and could make such long-range expeditions in a very short period of time. If our suspicions are correct, they may even have known something of our geological history, such as the occurrence of the 10,750-B.C.E. cataclysm that, legend tells us, brought a formerly flourishing global civilization to an abrupt end.

Yet one thing remains puzzling about the Crab Nebula ETI hypothesis, and this concerns the thousands of years involved between the creation of this celestial map and the time when its message could be seen by Earth observers. The Crab Nebula is located about 6,585 light-years from us. In real time, this supernova explosion would have occurred around 5530 B.C.E. and only some 6,585 years later, in 1054 C.E., would it have become visible on Earth. But in 5530 B.C.E. humanity was still in the Neolithic stone-age period. The Old Kingdom of ancient Egypt did not emerge until around 3100 B.C.E. Supposing the galactic community had been informed of humanity's relatively low level of advancement in that era, how could it be sure that 7500 years later humans would have progressed to the point where they had the technological means to study the Crab Nebula and its pulsar? Large optical telescopes and radio telescopes are a prerequisite for the Crab Nebula's unique features to be adequately discerned. Moreover, a sophisticated knowledge of both galactic and extragalactic astronomy is needed to understand the Crab Nebula's superwave warning message. The Crab Nebula sky map has been ready for us to study for at least the past 600 years, maybe even for the past 900 years. There are some centuries of leeway, then, as to the time when the prerequisite science and technology infrastructure would finally be developed and its message finally understood.

But if this is an intentional "time-capsule" message sent from the past, did its crafters have astounding precognitive abilities to know that we would be ready to receive it 6500 years later? In U.S. government research experiments, trained remote viewers have demonstrated up to

80 percent accuracy in viewing distant events taking place even thousands of miles away.[13] The same technique has been used to determine events that might occur in the future. Did alien clairvoyants remotely view our planet as it would be many millennia in the future? Or perhaps they knew that given the size of the human population and its rate of growth, the number of people would have grown sufficiently large to have spawned a civilization of adequate technological advancement some seven millennia later. We encounter the same conundrum in interpreting the one-radian message displayed by the millisecond pulsars residing in the Vulpecula/Sagitta region of the sky.

As before, we may want to consider one other possibility—namely, that members of this galactic ETI network have been working behind the scenes to influence the evolution of human civilization, thereby ensuring that we would be sufficiently advanced when their pulsar message was ready for viewing. Egyptologists have marveled at the sophisticated level of mathematics, science, engineering, and medicine that seemingly sprang out of nowhere at the dawn of civilization in ancient Egypt, much of which was progressively lost or forgotten over subsequent millennia.[14] Egyptian myth holds that the gods Osiris and Thoth (Hermes Trismegistus) had visited their land and taught humanity the arts of civilization and the Hermetic knowledge. Some authors have interpreted this literally as evidence of ETI contact.

Alternatively, perhaps these civilizations simply chose this particular time for their message to be viewed because they knew in advance that another superwave was due to pass through our part of the Galaxy around this time. Their efforts to warn us at this time may have been intended as a precautionary measure in case we had not figured things out on our own.

The Eclipsing Binary Millisecond (EBM) Pulsar exhibits a "coincidental" correspondence with the Vela pulsar period that is obvious only if their signals are observed during the past several decades. That is, the signal timing from the EBM pulsar shows a sinusoidal variation of ±0.089226 second over the course of its 9.2-hour orbital cycle. This corresponds to the time taken for the pulsar's radio signals to transit the 26,770-kilometer radius of the EBM Pulsar's *highly circular* orbit.[*][15]

---

*As the pulsar cycles farther from us and nearer to us, its pulsed signal alternately retards and advances by ±0.089226 second from the value it would normally have if it remained stationary. This interval of 0.089226 second represents the time required for the pulsar's radio signal to traverse the radius of its orbit about its companion star.

This signal timing variation happens to match very closely the pulsation period of the Vela pulsar, the discrepancy presently being only 0.09 percent! In fact, Vela's period had precisely this 0.089226-second value in September 1973 and has since been diverging from it by about 0.004 percent per year due to the gradual slowing of its pulsation rate. One thousand years before or after this date, Vela's period would depart from this orbital radius value by 4 percent, and the association would not be as obvious.

If these pulsars are entirely of natural origin, then we must consider it a simple stroke of luck that we are observing them at this unique point in time, during their many thousands of years of existence, when they would show this correspondence. If instead they are of ETI origin, their periods being carefully chosen to convey an association between the EBM Pulsar and the Vela pulsar, then some degree of foresight was necessary. Since the Vela pulsar, the nearer of the two, is located about 815 light-years from us, an alien civilization would have had to plan in advance at least 815 years for this period coincidence to be occurring at present. If the Vela pulsar dates back 12,000 years or more, then this planning would have had to span an even greater period.

The possibility that the EBM/Vela pulsar synchronicity was intentional becomes more plausible when we realize that it fits with other parts of the metaphorical pulsar message we have been analyzing. As we saw in chapter 2, the EBM Pulsar is the closest pulsar to Gamma Sagittae, which suggests an associative link with this one-radian indicator star. Also as we saw earlier, the $P–\dot{P}$ constellation map makes sense if we interpret the Vela pulsar coordinate as representing the star Gamma Sagittae. Thus, since the EBM and Vela pulsars are both associated in common with Gamma Sagittae, a symbolic association between these two pulsars seems appropriate.

Furthermore, by "encoding" the Vela pulsar period as the *radius* of the EBM Pulsar's highly circular orbit, the association of these two pulsars appears to be combined with the one-radian concept: that is, the 0.089226 light-second radial distance subtends an angle of one radian when laid out along the EBM Pulsar's nearly perfectly circular orbital circumference. The EBM Pulsar also portrays the one-radian concept in connection with Gamma Sagittae. As we saw earlier, this pulsar is so positioned in the sky as to form a right angle with Gamma Sagittae and with the Galaxy's nearby equatorial one-radian point (fig. 15).

Together, these allusions emphasize the idea of interpreting both Gamma Sagittae and the Vela pulsar as indicating a one-radian arrow flight from the Galactic center to the solar environs (see the text box below).

Is the Degree a Galactic Standard of Angular Measurement?

Dividing the orbital radius of the EBM Pulsar (0.0892267 light-second) by the period of the Millisecond Pulsar (0.001557806 second) yields the ratio 57.2772, a number that is within three hundredths of a percent of the number of degrees in one radian (57.2958). In this indirect fashion, these two pulsars could be expressing the one-radian concept. However, since the EBM pulsar's partner star is losing mass, its influence on the pulsar will be gradually decreasing, so the pulsar's orbital radius should also be gradually decreasing. At some time in the past, then, the pulsar's orbital radius in light-seconds would have been 29 micro light-seconds longer, making it large enough for this ratio to be numerically equal to the number of degrees in a radian. Knowing that the binary's orbital period is increasing at the rate of $3.8\pm10 \times 10^{-6}$ percent per year, we can estimate that the pulsar's orbital radius would have been 29 micro light-seconds longer $8,500\pm20,000$ years ago, or very roughly around the time of the Vela supernova explosion.

It is also interesting to note that if we take the period of the Vela pulsar and divide it by the period of the Millisecond Pulsar, we find that this period ratio was exactly equal to the number of degrees in one radian in September 1963 and has since been deviating from this value by ~0.002 percent per year due to the gradual increase of the Vela pulsar's period. We cannot be certain that a galactic commune of intercommunicating civilizations would be using a system of angular measurement that subdivides a circle into 360 degrees. Nevertheless, it is interesting to find that the period ratio of these two unique pulsars equals the number of degrees in one radian. This makes us wonder where our system of a 360° circle originated. Was it taught to us?

SEVEN

NATURAL OR ARTIFICIAL

Lighthouse Trouble

Supernova Origin Problem. Although astronomers continue to use the neutron star lighthouse model to interpret the pulsar phenomenon, they readily admit there are many things this model leaves unexplained. One shortcoming is that it fails to explain how pulsars originate. The lighthouse model predicts they should be formed in supernova explosions. However, there is a lack of evidence of pulsars residing within or near existing supernova remnants. Although 3 percent of the known pulsars have been found to coincide with remnants, a supernova cause remains uncertain, even for these few. In the case of the Vela and Vulpecula pulsars, the evidence contradicts the idea that they were formed by the explosion that produced their associated remnants. As was noted in chapter 5, the trajectory of the Vela pulsar is not aligned with the explosion center of the Vela supernova remnant. Nor could it have covered the distance from that site to its present location in the 13,000 years since the time of the Vela supernova explosion. Also, it is not certain that the Vela pulsar is located at the same distance as the Vela supernova remnant. By noting the frequency-dependent time lag of its radio signals, astronomers had estimated the pulsar should lie about 1,650 light-years away. But this is twice as far away as the estimated distance to the Vela remnant.

Astronomers concur that a causal connection between the Vulpecula pulsar and its remnant is doubtful as well because, judging from its large size, the remnant appears to be much too old in comparison with the pulsar. The only case where a pulsar can convincingly be shown to have originated from the center of a supernova explosion site is the Crab pulsar. But this does not prove that it is a spinning neutron star. The source may simply be the remnant core of the exploded progenitor star. The non-pulsing point X-ray source found inside the Cassiopeia A remnant close to its supernova explosion site could similarly be the hot remnant core of that exploded star as well.

Circular Orbit Problem. Pulsar astronomers have been surprised to find that a large percentage of pulsars are associated with an orbiting companion, for if pulsars are formed in supernova explosions, as their theory predicts, any nearby celestial bodies should have become propelled far from the explosion site and from the newly formed neutron star. For this reason, astronomers assume that a pulsar acquires its companions *after* the explosion through gravitational capture of closely passing bodies. In such a case, the captured planets or stars should have highly elliptical orbits much like short-period comets orbiting the Sun. But contrary to this expectation, most binary pulsars are orbited by bodies that follow exceedingly circular orbits. One example is pulsar PSR 1257+12, which is found to be associated with two planets, each 3 to 4 times the mass of the Earth, which follow highly circular orbits around the pulsar.[1] Another pulsar, PSR 0329+54, is orbited by a companion having an even lower mass, somewhere between 6 and 57 percent of the Earth's mass.[2] Again, the data indicate that its orbit is quite circular, posing serious problems for a supernova origin.

PSR 1957+20, the Eclipsing Binary Millisecond (EBM) Pulsar, also presents problems for the standard neutron star theory. Its companion is estimated to be a white dwarf with a mass equal to about 2 percent of the Sun's mass. Although this is considerably more massive than a planet, even a body this massive could not have survived the expulsive force of a supernova explosion. Pulsar timing indicates that this companion's orbit has an eccentricity of less than 20 parts per million, indicating that its orbit deviates from perfect circularity by less than one part per billion.[3] Compared with the Earth's orbit, this one is five thousand times more precisely circular. The circularity of its orbit precludes the possibility that the pulsar had gravitationally captured its white dwarf companion after the explosion. It is also unlikely that its companion spontaneously grew from material orbiting the pulsar, since all current neutron star theories suggest that the pulsar is eroding its companion, not contributing to its growth. The body orbiting another pulsar, PSR J2317+1439, exhibits an even smaller orbital eccentricity of just 0.5 part per million!

Orbits so *nearly perfect* in circularity are a rarity in astronomy. We find them only in connection with pulsars. Since no adequate explanation is forthcoming from theories proposing that pulsars originate in supernova explosions, this raises the question as to whether pulsars and their orbits may be one more aspect of ETI engineering. If pulsars have masses comparable to that of the Sun, as astronomical measurements seem to indicate, their creators, then, would have to possess technologies far beyond what we know in order to enable them to engineer bodies of such enormous mass. Those who want to stick with the theory that pulsars are natural objects such as spinning neutron stars must posit the repeated occurrence of a miracle of nature to explain these findings. In that case, a viable possibility is that we are just beginning to discover that nature is itself intelligent.

Multiple Pulse Problem. The lighthouse model also has difficulty accounting for pulse profiles that are composed of two pulses: a main pulse and an interpulse. In order for two pulses to be seen per cycle, the neutron star would have to emit two beams of synchrotron radiation, one from each of its magnetic poles, and the magnetic pole axis would have to be oriented nearly perpendicular to the neutron star's spin axis. Then, as the star rotated, an observer aligned with its equatorial plane would see pulses alternately coming from each radiation beam. In cases where the cosmic ray beams were not oriented close to the star's equator, the lighthouse model predicts viewers should instead see only a single pulse with each rotation.

However, this model has been unable to explain the pulses coming from the Crab Nebula pulsar. In this case, the interpulse follows the main pulse by 40 percent of a cycle, rather than by half of a cycle, as the dual-beam lighthouse model would predict (see fig. 41). To produce such a pulse pattern the pulsar's magnetic poles would have to be bent at an angle of about 145 degrees with respect to each other instead of directly opposing one another, as would normally be expected for the magnetic field of a celestial body.

The Crab pulsar is one of the few pulsars that emit pulses at the high-energy end of the electromagnetic spectrum. As mentioned earlier, besides radio pulses, it produces flashes of visible light, X-rays, and gamma rays. But the lighthouse model runs into trouble when it attempts to explain how pulses in these diverse portions of the spectrum might be generated from the same set of neutron star electron beams. Although the pulse profiles in all these spectral regions are synchronized with one another, in other respects they are substantially different. For example, the radio pulse profile is much narrower than the optical, X-ray, and gamma ray profiles (fig. 41). Also, the pulsar's signal intensity is found to drop off with increasing frequency at differing rates in each of these energy ranges, indicating that the radio, optical, X-ray, and gamma ray pulses are produced by different cosmic ray sources. This poses a serious problem for the standard lighthouse model, which hypothesizes a single beam of cosmic ray electrons coming from each magnetic pole.

Another feature that poses difficulty for the Crab pulsar lighthouse model is that the number of pulse components changes radically depending on which portion of the radio spectrum the pulsar is observed at. For example, between frequencies of about 430 to 606 megahertz, a small precursor pulse is found to immediately precede the main radio pulse, whereas at other radio frequencies, and also at optical, X-ray, and gamma ray wavelengths, only a single main pulse component is present. Moreover, the Crab pulsar's interpulse is present throughout the electromagnetic spectrum, from radio to gamma ray frequencies, but is *mysteriously absent* in the 2,700-megahertz-frequency range. To further complicate matters, at progressively higher radio frequencies in the 4,700 to 8,400 megahertz range, the interpulse returns *but is now shifted to an earlier phase,* now arriving 0.9 millisecond (2.8 percent of a pulse period) ahead of its previous position. *Moreover, the interpulse is now followed by two new peaks that did not show up in any other frequency range,* yielding a total of four peaks at 4,700 megahertz. But at 8,400 megahertz, *the main pulse peak disappears entirely,* leaving just the interpulse and the two additional peaks.

All of these peculiarities present enormous problems for scientists attempting to account for the Crab pulsar emission in terms of a natural phenomenon. The astronomers David Moffet and Timothy Hankins comment about this finding:

> In a multifrequency study of the Crab pulsar, we have found new and unusual components that defy explanation by emission from a simple dipole field geometry.[4]

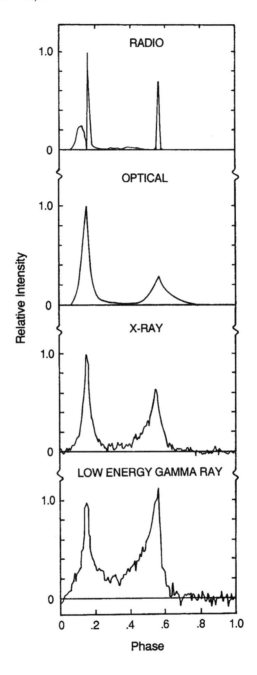

Figure 41. Time-averaged pulse profile for the Crab Nebula pulsar, as seen at radio, optical, X-ray, and gamma ray wavelengths (after Wilson and Fishman, Astrophysical Journal, *figure 4; Taylor and Manchester,* Annual Reviews of Astronomy and Astrophysics, *figure 4).*

The lighthouse model is confronted with an equally puzzling situation when attempting to explain the emission from the Vela pulsar. Like the Crab pulsar, Vela emits two pulses per cycle in the optical and gamma ray spectral regions. However, throughout the radio spectral region, it produces just one pulse per cycle (see fig. 42). Moreover, Vela's radio, optical, X-ray, and gamma ray pulses are not synchronized with one another. At gamma ray frequencies, its main pulse and interpulse are separated by 42 percent of a cycle, whereas at optical and X-ray frequencies they are separated by just 23 percent of a cycle. Its single radio frequency peak is also out of phase with all of these higher-energy peaks. So, as with the Crab pulsar, the dual-beam lighthouse model fails to account for the Vela pulsar's complex emission. Some theorists have been tempted to postulate three pairs of magnetic poles projecting particle beams from various points on the star's surface, but in the realm of natural phenomena this begins to approach the boundaries of the unbelievable. Models that attempt to explain the Crab pulsar's behavior fail to account for Vela's totally different pulsation characteristics.

Spin Axis Orientation Paradox. Pulsars that combine a wide pulse profile with a dual-pulse signal (main pulse plus interpulse) present a serious problem for the lighthouse model. For example, the Crab pulsar's pulses span 60 to 70 percent of its cycle. Vela's optical, X-ray, and gamma ray pulses span 70 to 80 percent of its pulse cycle. Nevertheless, both pulsars produce bimodal pulse profiles that have both a main pulse and an interpulse. The lighthouse model attempts to explain such long-duration emission by assuming the pulsar's rotation axis is aimed almost directly toward the observer, so its radiation beam never disappears entirely from view as the neutron star rotates. However, to explain the interpulse phenomena lighthouse model theorists make the contrary assumption that the neutron star's rotation axis is instead aimed almost *perpendicular* to our line of sight. In this way, oppositely directed beams would alternately sweep in our direction. So far no one has been able to resolve this spin axis orientation conundrum.

The pulsar PSR 0950+08 also poses a problem. Although this pulsar is seen to produce a distinct main pulse, close observation shows that it produces synchrotron radio emission *over its entire pulse cycle*. Thus, according to the lighthouse model, the neutron star's rotation axis should be aimed almost directly toward us, so we should be seeing only one of the pulsar's two radiation beams; the other should remain

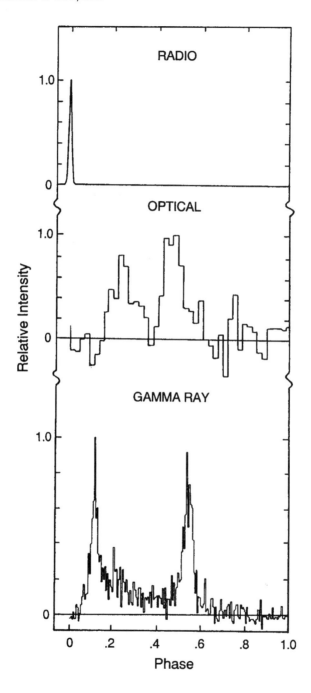

Figure 42. Time-averaged pulse profile for the Vela pulsar, as seen at radio, optical, and gamma ray wavelengths (after Kanbach et al., Astronomy and Astrophysics, *figures 1 and 2).*

hidden from our view. However, contrary to expectation, an interpulse is seen to follow its main radio pulse by about 42 percent of a pulse cycle (see fig. 5), implying that we are seeing a second beam oriented 150 degrees to the main beam.[5] This paradox has yet to be resolved.

Another perplexing feature of PSR 0950+08 is that its interpulse shows up only in a few percent of the pulse cycles. If this pulsar is a spinning neutron star, why would it be producing only one pulse the other 98 percent of the time? The lighthouse model offers little in the way of an explanation. As a result, together with the Crab and Vela pulsars, PSR 0950+08 stands as *one of the strongest pieces of obser-vational evidence refuting the neutron star lighthouse model.* If these pulsars are interstellar communication beacons, then the civilizations who created them should be commended for their ingenuity, for their signals have thrown astrophysicists and their models into a tailspin.

Uniform Emission of Cosmic Rays. PSR J0437–4715, the second closest millisecond pulsar to our solar system, lies just 450 light-years away in the constellation of Pictor.[6] Because of its close proximity, astronomers have been able to obtain reasonably detailed radio tele-scope images of it. These show that it is closely orbited by a white dwarf star and that both are surrounded by a bow-shaped shock front (fig. 43) similar to that surrounding the EBM Pulsar (fig. 16). Astronomers interpret this front as the boundary where the pulsar's cosmic ray wind confronts the surrounding gaseous interstellar medium as the pulsar plows forward. But the symmetrical shape of the bow front has led them to conclude that this pulsar, like the EBM Pulsar, *must emit its particle wind with equal intensity in all directions.* This, however, contradicts the theory that a pulsar is a neutron star that emits a pair of magnetically confined cosmic ray particle beams from either of its poles. If pulsars emit their cosmic ray particles evenly in all directions, how would their rotation produce brief radio pulses? The lighthouse model does not provide an answer.

Relativistic Rotation Problem. The spinning neutron star model was stretched close to its limit of credibility with the discoveries of the Millisecond Pulsar, PSR 1937+214, and the EBM Pulsar, PSR 1957+20, which spin respectively at 642 and 622 revolutions per second. The light-house model proposes these to be 20-kilometer-diameter neutron stars spinning so fast that they have surface velocities of about *13 percent of the speed of light.* This is very close to the limiting velocity beyond which a neutron star would be made to fly apart. If the neutron star were indeed spinning at such a high speed, the forces generated by its rotation would be expected to continuously agitate the neutron star's magnetic field and cause large variations in the shape of the pulsar's time-averaged pulse profile. This problem is expected to be most severe in the case of the EBM Pulsar since its companion, which is separated from it by just one solar diameter, is obviously interacting quite strongly with the pulsar, being brightly illuminated by the pulsar's ionizing particle wind. However, as with most other pulsars, the pulse envelopes of these millisecond pulsars have a characteristic shape that does not change over years of observation. In fact, the EBM Pulsar produces a main peak that has a duration of just 35 microseconds, 2 percent of its pulse period, which makes its profile one of the most precisely defined of all pulse profiles. The light-

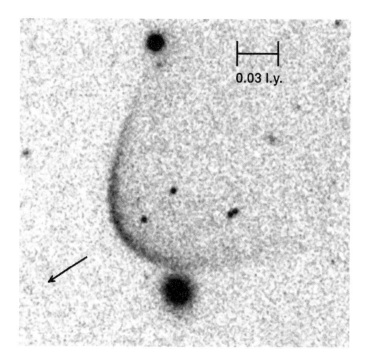

Figure 43. Optical image of the binary millisecond pulsar PSR J0437–4715 and its surrounding bow shock front taken through a red filter (15" = 0.03 light-year). The white dwarf companion is visible as the faint star directly behind the shock. The pulsar lies at that same location but is not optically visible. The arrow shows the direction in which the pulsar and its companion are moving in the plane of the sky (courtesy of Andrew Fruchter of STSI and the Cerro Tololo Inter-American Observatory).

house model fails to account for the pulse profile constancy under such high-speed conditions. This difficulty, however, is readily resolved if pulsars are theorized to be non-rotating communication beacons set up by alien civilizations.

Period-constancy Problem. Another unusual feature of millisecond pulsars is the extremely low rates at which their periods change. According to the classical lighthouse model, such rapidly spinning pulsars should have very high rates of energy loss through the production of gravity waves and electromagnetic radiation and hence should be slowing down rapidly. Calculations predict that the Millisecond Pulsar and the EBM Pulsar, which presently have speeds of more than 600 revolutions per second, should have slowed to about 100 revolutions per second in just one year.[7] Hence, astronomers expected them to have the most rapid rate of pulse period lengthening of any pulsar, almost 1000 times that of the Crab pulsar. But just the opposite is found. The period of the EBM Pulsar lengthens at the extremely slow rate of just 1.7×10^{-20} seconds per second, or about half a picosecond per year. This is about 25 million times slower than the Crab pulsar's retardation rate. The Millisecond Pulsar has a comparably slow rate of change of just 1.05×10^{-19} seconds per second (3.3 picoseconds per year). In fact, the 20 most constant pulsars in the sky are all millisecond pulsars and thus behave contrary to theorists' expectations.

Astronomers also had supposed that pulsars with high spin rates would be relatively young, with millisecond pulsars being at most a few years old. If so, then they should be associated with a recent supernova explosion. But none of the 120 or so millisecond pulsars discovered thus far has been found to be anywhere near a young supernova remnant. Thus their discovery has proved to be quite troublesome for astrophysicists.

Spin Energy Dissipation Problem. Lighthouse model theorists believe the explosive forces involved in compressing matter to neutron star densities would also compress and amplify any preexisting magnetic fields to very high values. They presume that a neutron star pulsar would have a very strong magnetic field. Moreover, their calculations indicate that as the neutron star spins, its rotating magnetic field will experience a dynamo braking force that will convert some of the star's rotational kinetic energy into electromagnetic radiation. As a result, the neutron star would progressively slow down, and its pulse period would gradually lengthen. But considering that pulsars have masses comparable to that of the Sun, they would have to be losing very large amounts of kinetic energy to accommodate their slow-down rates. Lighthouse modelers, then, are forced to assume that their neutron stars must be radiating energy away at an enormous rate. To obtain an adequate braking force, they assume that the neutron star has an excessively high magnetic field strength, ranging from a billion to one thousand trillion times the intensity of the Earth's magnetic field. This *assumed* magnetic field strength is about a billion times greater than field strengths normally observed in compact stars such as white dwarfs. But the main problem is that the power that must be radiated as a result of this hypothetical braking force is calculated to exceed by at least a thousand times the amount of power being radiated in the pulsar's beamed emission.[8]

So where is the remaining 99.9 percent of the energy the lighthouse predicts pulsars should be dissipating? If pulsars radiated this in the form of electromagnetic radiation, our radio telescopes should have seen it, and it should have been so intense as to mask the pulsar signals. But none is seen. If this missing energy instead went into accelerating cosmic rays, the acceleration process should produce large amounts of electromagnetic emission and again we should see something. But we don't.

The Crab pulsar presents an extreme example of this problem of the missing emission. If the cosmic rays illuminating the Crab remnant were being accelerated by the pulsar's spinning magnetic pole, the acceleration process should produce an electromagnetic emission equal to that coming from the entire remnant, but concentrated within a million-kilometer-diameter region centered on the pulsar. Our radio telescopes, which can resolve radio sources to an angle of at least one milliarc second, should then detect an unpulsed compact radio source having an intensity *100 billion times greater than that of the diffuse background emission from the Crab remnant, and over a trillion times greater than the pulsed radio signal coming from the Crab pulsar.* But a radio source of such high intensity is not seen within the remnant.

The answer to this problem is simple. There are no "spinning neutron stars," rapidly dissipating their rotational energy. Pulsar radiation beams are stationary, and their pulse rates are precisely controlled by intelligent design, not by any phenomenon of nature. As for the Crab Nebula, the cosmic rays that produce its diffuse emission originate from outside the remnant, not internally from a neutron star source.

Signal Ordering
Too Complex to Explain

If a civilization wanted to initiate communication with a civilization elsewhere in the Galaxy, it would want to send a signal that would not be mistakenly interpreted as coming from a natural object. Because most naturally occurring astronomical phenomena produce emissions that have a large degree of irregularity, a communicating civilization would want to make sure that its transmission was instead very highly ordered. In this respect, of all galactic emission phenomena, pulsars seem to best fit the criterion of being of artificial origin. No other type of radiation source has been found to exhibit such precise time ordering. Regular periodic variations in luminosity and color are known to occur in variable stars such as Cepheid variables, Mira variables, pulsating white dwarfs, and X-ray stars, but their pulsation periods are not nearly as accurate as those seen in pulsars. Some binary X-ray stars, for example, are known to pulse with periods as short as a few seconds, but their periods are accurate only to six or seven significant figures.* Pulsar periods, on the other hand, typically range from a million to 100 billion times more precise. In addition, pulsating X-ray stars do not exhibit the rich variety of ordered behavior seen in the signals of radio pulsars. In instances where radio continuum emission is detected from such X-ray stars, the emission is usually highly erratic, as is characteristic of stellar flaring. Such pulsating sources may truly be categorized as being of natural origin. Pulsars, on the other hand, are in an entirely different category. Their pulsation patterns are so highly ordered and so incredibly complex that attempts to describe them in terms of natural processes border on absurdity.

Let us review some of the ordering characteristics of pulsar signals. As was noted in chapter 1, the lighthouse model has difficulty accounting for even the most basic of pulsar signal characteristics, such as explaining how the time-averaged pulse period is so precisely timed in view of the unusually large timing variability that characterizes the individual pulses. As we have seen above, the lighthouse model also has trouble explaining the period derivative characteristic, the gradual slowing of the pulsar's rate of pulsation. In addition, however, there is a

---

*Hercules X-1, for example, has a period of 1.237772±0.000001 seconds and Centaurus X-3 has a period of 4.843496±0.000007 seconds.

great depth to the complexity of pulsar signals that poses equally grave difficulties for theories attempting to explain pulsars as being of natural origin. The great variety of signal ordering that is observed in pulsars is shown in table 4 and treated in some detail in appendix A. Some main points are summarized in the remainder of this section.

Pulse Amplitude Modulation. Although pulse timing and pulse amplitude change considerably from one pulse to the next, this variation is often found to occur in an ordered fashion. For example, in some pulsars the pulse intensity is observed to vary in a regularly timed manner, brightening for one or two pulses out of a long sequence of pulses. This pulse modulation typically occurs from 2 to 20 times slower than the repetition rate of the pulsar's *primary* pulse cycle.

To understand this layered ordering, imagine that you are looking through a rotating strobe wheel at a fluorescent light on a room ceiling. By rotating the strobe wheel at the proper speed and looking through its slits as they flash by your eye, you are able to match the frequency of the light as it flashes on and off 60 times per second. At this strobe speed, the flashing of the light will appear to you to have "frozen" in time. Depending on the phasing of the strobe relative to the light, the light will be seen to be either steadily illuminated at full brightness, steadily illuminated at semi-brightness, or steadily dark. Now, suppose something has gone wrong with the central power-generating station and that this causes the lights to momentarily brighten about five times every second. When we look through our strobe at the fluorescent light, we now see that its brightness is no longer constant, but rather it flickers at a frequency of five Hertz. The 0.2-second period of this *amplitude modulation* would be found to be 12 times as long as the lamp's primary period, which has a one-sixtieth-second duration. This secondary amplitude modulation would be analogous to the pulse amplitude modulation we are talking about.

This amplitude modulation of the pulsar signal is also analogous to techniques commonly used in AM radio broadcasting. In this respect, the primary pulse repetition rate would be somewhat like the radio station's carrier frequency and the amplitude modulation of the pulsar's signal might be comparable to the amplitude modulation imposed on this carrier to broadcast music.

But pulsars can exhibit even more complex ordering, particularly those that have a wide pulse profile. For example, in some pulsars the

TABLE 4. KINDS OF ORDERING OBSERVED IN PULSAR SIGNALS

1. The 10-to-17-decimal-place precise regularity of the pulsar's primary period, P_1, as determined from the timing of its time-averaged pulse profile.

2. The 6-decimal-place precise regularity of the rate, $\dot{P}_1$, at which the pulsar's primary period increases over time.

3. The constancy of the shape and polarization of the pulsar's time-averaged pulse profile.

4. The ordering of successive pulses occurring within the same pulse cycle such that the pulses are always spaced from one another by a characteristic time interval, P_2.

5. The amplitude modulation of a sequence of consecutive pulses with a characteristic modulation period, P_3.

6. The occurrence of many pulse modulation frequencies in one pulse cycle with the modulation frequency being ordered with respect to pulse cycle phase. That is, a specific modulation frequency occurs only over a specific phase range.

7. The correlation of sequential pulse occurrences where a pulse occurring at one phase of the pulse cycle will be followed by a pulse occurring at a later phase of that pulse cycle.

8. The phenomenon of pulse drifting, whereby pulse sequences repeatedly scan across the time-averaged pulse profile with a characteristic repetition period, P_3.

9. Pulse "freezing" during nulling periods: the instance when the pulse signal becomes undetectable and pulse drifting ceases during an interval of time, after which drifting begins again where it left off.

10. Rapid amplitude modulation of a pulse's intensity, with the characteristic period, P_4, being on the order of tenths of a millisecond.

11. Mode switching, when the pulsar abruptly switches from one stable organized signal pattern (pulse profile) to one or more other stable organized patterns (pulse profiles).

12. The quantization of pulse drift rates, where the pulse drift rate doubles or triples in whole-number multiples when the pulsar switches from one mode to the next.

13. Frequency dependent mode switching, where the pulsar's mode switching is apparent only when the pulsar signal is observed within a certain radio frequency range.

14. Mode switching grammar, where mode switching is governed by rules that determine what mode the pulsar may switch to from any given mode that it happens to be pulsing in.

period of this amplitude modulation can differ depending on the phase of the pulse cycle. For example, one might find that pulses appear 4 times per second near the beginning of the time-averaged pulse profile, 7 times per second near the end of the profile, and that they recur seemingly randomly in the middle part of the profile. Further complicating matters, two or more pulse modulation periods may be found to be present at any given phase of this pulse cycle. For example, one might find that pulses recur both 4 and 5 times per second near the beginning of the time-averaged pulse profile.

As yet another type of ordering, pulse modulation at one phase of the pulse cycle might be found to correlate with pulse modulation at another phase of the pulse cycle, but with a built-in time lag so that a pulse occurring at one phase of the pulse cycle would be followed one pulse cycle later by a pulse occurring at another phase of the pulse cycle.

No natural stellar source is known to exhibit such highly organized signal characteristics, not even just one of these various types of pulse modulation. Astronomers seem to have taken for granted the idea that a natural object could conceivably produce such ordering, since they have presumed that the unusual nature of this ordering might be attributed to the exceptionally high degree of matter ordering that is supposed to exist within a neutron star. Thus, as they catalog these various ordering characteristics, they think they are investigating the unusual conditions of ordering they suspect might exist within a neutron star. Actually, they have no independent astronomical observations to confirm the assumptions they are making. When asked for proof of the existence of neutron stars, all they can point to are the pulsar observations, the very phenomenon they are attempting to explain in terms of their neutron star model. Nor have they conducted any experiments in the laboratory to determine if a naturally rotating particle beam could in fact generate such signal ordering. In the way of explanation, they confine themselves merely to the task of "modeling" a phenomenon they a priori *presume* to be natural.

Pulse Drifting. Pulse drifting is a form of pulse amplitude modulation in which the pulse appears to "drift" across the time-averaged pulse profile. In other words, its phase in the pulse cycle progressively changes with each successive pulse so that its position moves from the leading edge toward the trailing edge of the pulse profile, or in the reverse direction. As with pulse modulation, this pulse drifting phenomenon does not occur in a simple fashion. Often drifting is restricted to certain pulse cycle phases, with drifting occurring at different rates depending on which part of the pulse cycle one happens to be analyzing.

In addition, there are various types of drifting behavior. There is *linear drifting,* in which the pulse scans across the time-averaged pulse profile at a constant rate. There is also *nonlinear drifting,* in which the pulse drift rate changes, possibly beginning at a slow rate, gradually accelerating to a high drift rate, then gradually decelerating to a slow drift rate. Again, just as the shape and timing of the time-averaged pulse profile remain exceedingly constant over time, so too these various specific characteristics of pulse modulation and pulse drifting do not vary from one year to the next. They are somehow programmed into each pulsar, providing each pulsar with a set of unique and unchanging signal characteristics.

If pulsar signals are of artificial origin, their authors have apparently done an excellent job of achieving a key objective in designing the ideal interstellar communication—namely, they have made the ordering of their transmissions so numerous as to confound astronomers' best attempts to ascribe the signals to a natural cause.

Pulse Nulling and Freezing. Some pulsars produce a temporary signal blackout. One second they are transmitting at full intensity and the next second their signal has vanished. These so-called *null periods* have been found to last for 8 hours or more. Careful studies of a few pulsars that exhibit this phenomenon show that during nulling the pulsar is still transmitting pulses, but at a very low intensity. Observations have also shown that during nulling the pulse drift rate becomes exceedingly slow. Consequently, when normal transmission resumes and the pulse drifting recommences at its normal rate, the pulses continue to scan the profile almost from the same pulse cycle phase they were in when they disappeared from view. Pulse nulling may be compared to the screen-saver or energy-saver feature on a home computer that lowers screen luminosity as a way of reducing the rate of energy consumption and extending the lifetime of the computer's cathode-ray tube. The pulsar's freezing feature conveniently retards the pulse drifting activity during nulling with the result that little is missed during the interval in which nulling is in progress. It is as if the pulsar "remembers" which phase of the pulse cycle its pulse had appeared in at the time nulling began. Theorists have met with considerable difficulty in their attempts to explain this nulling and pulse freezing phenomenon, which must be included as one more ad hoc addition to their lighthouse model.

Mode Switching. Of the various pulsar signal-ordering phenomena, mode switching is probably one of the most interesting and most complex. Whereas most pulsars have one stable pulsation mode that persists seemingly indefinitely, in mode-switching pulsars, the normal pulsation mode might suddenly vanish and be replaced by a different and equally complex alternate mode. Whereas a particular mode of pulse ordering may last anywhere from 10 to 10,000 or so pulse periods, the transition from one mode to another may occur in as short a time as one pulse period. A mode-switching event changes the entire character of the pulsar signal. Ordered patterns that normally remain remarkably invariant over many pulse periods suddenly terminate and a new invariant pattern takes over. Mode switching alters the shape of the time-averaged pulse profile, restructures the pulse-drifting and modulation behavior, alters the pulsar's polarization properties, and changes the way in which the pulse profile component intensities vary as a function of radio frequency.

To draw an analogy, a pulsar's time-averaged pulse profile might be imagined as being built up from its sequence of pulses in somewhat the same way that a fax machine's facsimile image is built up from its series of transmitted dots. Mode switching would be equivalent to the instance in which an entirely new fax image, or page, begins to be transmitted. In spite of the radical changes involved in switching from one mode of signal ordering to another, the pulsar's exceedingly precise primary period and period derivative remain unchanged. This perplexing behavior has stymied astronomers' attempts to come up with a natural explanation. In 1982, one group of pulsar investigators wrote: "Despite the fact that the mode-switching phenomenon has such an important effect on pulse emission, no elaborate theory as yet exists as to how to interpret it."[9]

Frequency Dependent Mode Switching. In some pulsars the time-averaged pulse profile has the same form regardless of what frequency the pulsar is observed at. In other pulsars, the shape of the time-averaged pulse profile changes radically with observing frequency, as does the number of components making up its profile. This is particularly apparent in the case of the Crab pulsar, as was noted earlier. This frequency-dependent effect becomes all the more mysterious in the case of mode-switching pulsars. For example, in the case of the mode-switching pulsar PSR 0329+54, not only does the profile have differing shapes at different observing frequencies, but it also has differing switching modes available to it at different frequencies.[10] When observed at frequencies ranging from 14.8 gigahertz* down to 5 gigahertz, the pulsar is found to switch between its normal mode and a single *abnormal* mode designated as mode A (see fig. A.6 of appendix A). At a frequency of 2.7 gigahertz, the pulsar is instead found to switch between its normal mode and either of two abnormal modes: mode A and a new mode designated as mode B. At a frequency of 1.4 gigahertz, switching is found to occur between the normal mode and *three* abnormal modes: A, B, and C. Proceeding to still lower frequencies the number of available modes declines again, only abnormal mode, B, being available at 0.83 gigahertz and only abnormal mode A being available at 0.41 gigahertz. Lighthouse models have not provided an adequate explanation for these unusual frequency-dependent switching rules.

We are given to wonder whether there may be an intended logic behind the complex switching characteristics of this pulsar. That is, the pulsar seems to draw our attention to the 1.4 gigahertz frequency (21-centimeter wavelength) as being "more significant" than the rest since all three abnormal modes are available at that frequency and at no other. Interestingly, this is none other than the neutral hydrogen emission line frequency, the same frequency that SETI astronomers believe would most likely be used by alien civilizations for galactic communication.

Mode-switching Grammar. In cases where pulsars have several modes available to them, switching from one mode to another sometimes conforms to certain fixed rules. One such pulsar, PSR 0031–07, can adopt any of three possible signal-ordering modes: A, B, and C.[11] This pulsar emits bursts of pulses that are separated by null periods. During such a burst, rules come into play that govern how mode switching takes place. For example, if the pulsar is in mode A, it may switch to mode B within the same burst; if it is in mode B, it may switch to mode C within the same burst. However, the pulsar has never been found to switch from mode A to mode C in the same burst. Rules also appear to govern the rate at which pulses drift when each of these

---

*One gigahertz equals 1,000 megahertz.

modes is dominant, modes A, B, and C having pulse drift rates that are whole-number multiples of one another in the ratio of 1:2:3.

This quantized drifting behavior and mode-switching grammar poses a serious problem to theorists who attempt to explain it as a natural phenomenon. Subatomic particles in the nucleus of an atom are known to be restricted to specific quantum energy states with specific rules governing their transitions from one energy state to another. Theorists have speculated that the neutrons composing a neutron star may reside in similar quantum states due to their highly compact condition. But to say that the same highly organized behavior should apply to the cosmic ray beam emitted from the surface of a hypothetical neutron star requires a very large leap of inference. Nuclear quantum ordering governs particles that are bound together by strong force fields, whereas the cosmic ray particles that are theorized to be generating the lighthouse model radiation beam are instead in high-speed flight away from the star and hence would not necessarily reside in the same high-density region. Moreover, if we are to attribute pulse quantization and mode-changing grammar to neutron star ordering, why is it not more prevalent among pulsars, but rather is unique to this particular pulsar?

Modeling Difficulties. The lighthouse model theorists have been quite at a loss to explain the numerous types of highly ordered phenomena that pulsars exhibit. Consider, for example, the question of pulsar memory. How does a mode-switching pulsar "remember" its former pulsation mode so that it can later switch back to it? Does the cosmic ray electron beam interact nonlinearly with the neutron star magnetic field so as to have various semi-stable *chaotic attractor* states, each corresponding to a particular pulsation mode? And how does a pulse in a drifting sequence recall the phase of the preceding pulse so that it can become properly timed in the pulsation cycle and reproduce a drift velocity of a specific magnitude? And how are such drift rates maintained over thousands of pulse sequences? Astronomers can only speculate, since they have no laboratory data to back up this sort of behavior other than the pulsars themselves.

Some theorists have suggested models in which the pulse-drifting phenomenon is due to cosmic ray discharges, or *sparks,* precessing about the neutron star's magnetic field pole. But theorists acknowledge that due to the turbulent polar field environment, such precessing could be stable only for very short periods of time. Finally, how does a given pulsar modulate the phase and intensity of its pulses in such a way that the resulting time-averaged pulse profile has a specific shape and specific light polarization characteristics that remain invariant, or "remembered," throughout years of observation?

The pulsar astronomers Alexei Filippenko and V. Radhakrishnan have been troubled by such considerations. In 1982 they wrote:

> The standard polar cap [lighthouse] model of pulsar radio emission provides acceptable explanations for a wide variety of observed pulsar characteristics. Nevertheless, we show that it has difficulty accounting for certain details pertaining to drifting subpulses, nulling, and mode changing. In particular, the persistence of drifting subpulse phase memory observed during pulsar nulling, as well as the phenomenon of nulling itself, seem to defy simple explanation.[12]

A number of qualitative modifications of the lighthouse model have been tentatively proposed in an attempt to account for these special types of ordering. However, so far no one has been able to bring these disparate revisions together into one coherent model. The pulsar astronomers Joseph Taylor, Richard Manchester, and G. Huguenin have also expressed concerns about pulsar models:

> There exists an extensive literature of observational details which, although well substantiated by repeated measurements, remain unexplained and unassimilated into pulsar models.[13]

A theoretical model such as the lighthouse model not only provides astronomers with a framework for interpreting and comparing the various sets of observational data they come up with, but it also provides them with a common agreed-upon context for sharing their ideas with one another, either in conversation or through publication. As a result, it serves a practical purpose. However, because of the sizable mental investment involved, astronomers hesitate to discard their model, even if it fails to account for observation. The lighthouse model, therefore, has evolved into a theoretical paradigm, a conceptual system having the unintended properties of self-stabilization and self-preservation. When new types of pulsar-signal ordering are discovered that do not fit the lighthouse model, rather than seeking an alternative explanation, astronomers tend to elaborate their model or make special-case modifications so that it might accommodate their new findings. Consequently, lighthouse models have grown unusually complex and varied, leaving pulsar astrophysics in a position similar to that of pre-Copernican astronomy, which taught that the planets orbited the Earth in complex epicycle paths.

Each pulsar's signal is found to be highly ordered in a number of ways, and the manner of this ordering differs considerably from one pulsar to another, making it impossible to conceive of a single neutron star model capable of encompassing the great variety of observed properties. On the other hand, such varied ordered complexity is exactly what one would expect if pulsars were in fact beacons used for interstellar communication or space navigation.

It is often said that one's first hunch is usually the right one. Few of the pulsar-ordering properties that we know of today were known back in 1968 when the group of Cambridge University astronomers first published their pulsar discovery. At that time they knew just that the shape and pulse period of the time-averaged pulse profile remained highly invariant. If they had known then what we now know about pulsars, perhaps they would not have rejected the ETI communication scenario as readily as they did.

A "Low-tech" Particle-Beam Communicator

Using presently available technology, it should be possible to construct a device capable of transmitting broadband interstellar signals much like those coming from pulsars. The technology required is the same as that used in particle accelerators employed by high-energy physicists for carrying out particle-collision experiments. It is also the same technology used in particle-beam weapons systems such as those developed in the Pentagon's Star Wars program—except in this case the technology would be employed for peaceful purposes.

As mentioned earlier, a space-based particle-beam communicator would consist of two main components: 1) a high-energy particle accelerator and 2) a magnetic modulator unit for decelerating the particles and converting their kinetic energy into a beam of broadband synchrotron radiation (fig. 1). The accelerator module would accelerate electrons very close to the speed of light (differing from light speed by only a few parts in 100 billion). At this speed they would have energies of about 100 billion electron volts, about 200,000 times greater than their rest-mass energy.

By comparison, the Stanford University linear electron accelerator used for carrying out high-energy particle-collision experiments achieves electron energies of 50 billion electron volts. This device accelerates its particles by discharging electric power stored in enormous capacitor banks into metal collars positioned along the length of the accelerator tunnel. The steep electric field gradient induced within each collar gives the stream of particles a series of kicks that cause them to move forward at ever increasing speeds. The particles are then focused into a beam by a series of superconducting electromagnet rings positioned at intervals along the tunnel, magnets so powerful that they require special spring-loaded restraints to prevent them from exploding when powered up.

Another type of particle accelerator has recently been developed that promises to vastly extend the present state of the art. This so-called *beat-wave plasma particle accelerator* is able to generate accelerating energies per centimeter of tunnel length that are from 10,000 to 10 million times stronger than those used in the Stanford accelerator.[14] Two powerful laser beams of slightly differing frequencies are projected into a gas plasma tube to produce a "beat frequency wave" that moves through the plasma at a tremendous speed. Electrons "surfing" on this wave are then accelerated as the wave travels down the tube. A 10-meter-long accelerator of this sort is theoretically capable of accelerating electrons to 100 billion electron volt energies.

Turning now to the particle-beam modulator, which would be fixed at the output of the particle accelerator unit, this would consist of a long tunnel flanked by a series of powerful superconducting electromagnet coils. Their poles would alternate between north and south so that when the coils were energized, the passing electrons would be deflected transversely, first one way and then the other, all the while maintaining their original straight-line course down the tunnel (fig. 1). The electrons in the beam would emit broadband synchrotron radiation with each deflection. Due to their near speed-of-light velocity, they would project this radiation forward as *a very narrow beam*. Because the electrons would be traveling just about as fast as the beamed radiation, the emission contributed by each consecutive magnetic deflection along the length of the modulator tunnel would add up in phase to produce an extremely powerful, laserlike beam.

By pulsing the electromagnetic modulator at regular intervals, the particle beam could be made to produce periodic bursts of synchrotron radiation. The bursts could be made to occur at extremely regular intervals simply by controlling the intensity of the beam modulator with a control unit timed by an atomic clock. Depending on how it was programmed, the control unit could generate the various pulse patterns and ordering phenomena observed in pulsars. It could produce a mode-switching effect simply by flipping between alternate pulse contouring routines stored in its memory.

The energy of the beamed electrons and the strength of the modulator's magnetic field would together determine what part of the electromagnetic spectrum the synchrotron beam would most strongly radiate in. A communicator beacon using a bank of several particle accelerators, each producing an electron beam in a certain energy range and each having its own electron beam modulator, could be made to simultaneously project radio, optical, X-ray, and gamma ray pulses, producing pulse characteristics such as those seen in the Crab and Vela pulsars. Alternatively, a similar result might be accomplished by using a single particle beam with a multistage modulator that would apply deflecting fields of various strengths. With progressively stronger deflecting forces, the electron beam could be made to emit in progressively higher energy parts of the spectrum: optical, X-ray, or gamma ray.

Powered by a reasonably sized power source, such a particle-beam communication device could transmit a broadband radio signal across a distance of several thousand light-years and still yield a signal strength comparable to that detected from pulsars. It could even produce signals as intense as those coming from the Crab pulsar, one of the brightest pulsars in the heavens. At a distance of 6,600 light-years, a one-megawatt synchrotron radiation beam projected from a 100-billion-electron-volt electron beam would yield a radio-signal intensity comparable to that observed from the Crab pulsar, which is situated at a similar distance (see appendix B). Although a radio power output of one megawatt is about ten times the power radiated by some of the larger commercial radio stations, it is still an easily achievable power level. By comparison, some of the larger nuclear power plants are able to produce over 1000 megawatts of power.

The power of the beamed synchrotron radiation is proportional to the cube of the particle energy. Hence, a 100,000-fold increase in particle energy translates into a 100,000-fold-narrower synchrotron beam and into a 10^{15}-fold-higher beamed radio signal intensity. The pulsed communicator radiation beam could easily be aimed so as to be visible to Earth observers. For example, if the beam was confined to a cone angle of just one second of arc, at a distance of 6,600 light-years, it would be visible from a region 0.03 light-year across (1000 times the diameter of the Earth's orbit). Star-tracking devices such as those used to aim the Hubble Space Telescope could target the beam with an error of less than 1 percent of the beam's angular diameter.

The particle-beam-communicator model, described above and pictured in figure 1, is one initially devised when I began suspecting that pulsars might be ETI communication beacons. For many years I thought this was the only possible way to explain pulsars as an ETI phenomenon. At that time I thought pulsar signals might be coming from communication beams projected from specially built space-station installations. However, new data on binary pulsars that came out in the early 1990s forced me to abandon this design and confront the fact that pulsar signals must actually be coming from near the surfaces of very massive bodies. Such was implied from the cyclic variations in pulse period observed coming from pulsars that were in binary association with a companion star. By studying these variations, it is possible to determine not only the circularity of the companion's orbit, but also the companion's approximate mass and the mass of the pulsar. Invariably, the pulsar's mass turns out to be quite large, comparable to that of our Sun.

Thus, I realized that pulsar signals must necessarily be emitted in the rest frame of a very massive star, and cannot originate from the kind of space-station outpost described above. Although such a communicator design could be used to produce beamed signals very much like those coming from pulsars, it could not account for the observation that pulsar signals come from bodies of celestial size.

Not having an adequate model, I began to consider the possibility that pulsars might be a natural phenomenon and that their conveyed message might reflect the presence of a high intelligence permeating the universe and attempting to make its presence known to us on a grand scale. However, shortly thereafter I thought of a different method of generating pulsar signals that made it possible once again to reconsider the hypothesis that pulsars might be artifacts of galactic civilizations. Using this method, an advanced civilization might actually engineer the cosmic ray flux coming from a star and cause it to emit a pulsed, collimated beam of synchrotron radiation. With sufficiently advanced technology, this should be possible. Let's examine one method.

Field-engineered Stellar
Cores as ETI Beacons

The first step in creating a pulsar communicator beacon would be to locate a suitable natural source of cosmic ray electrons. A stellar core left behind in the latter stages of a star's evolution would be quite suitable. That is, as a star matures over a span of billions of years, its dense, metal-rich core would grow increasingly massive and the energy flux from this core would become increasingly intense.[15] Eventually, the outpouring of energy from its core would be so great as to eject the star's gaseous envelope. This would leave behind an exceedingly hot, dense stellar core about the size of our Earth but having a mass anywhere from a few tenths to somewhat more than one solar mass. The core would be so hot that it would emit its energy in the form of high-energy cosmic ray particles rather than as visible light. In some cases, where a stellar core was still surrounded by a residual atmosphere, its cosmic ray flux would excite visible, ultraviolet, and X-ray radiation. As such it would be classified as a white dwarf star. In other cases, where the core's atmosphere would be more rarefied, the cosmic rays would be able to stream freely outward, exciting some X-ray and gamma ray radiation but little visible light. In such a case, the core would be classified as an X-ray star. In cases where it was bare of an atmosphere, it would emit cosmic rays accompanied by very little electromagnetic radiation and would be essentially invisible to detection. Unlike in the neutron star model that has the cosmic rays streaming out as confined beams, such bare stellar cores would radiate their cosmic rays fairly uniformly in all directions.*

In the case where a star had undergone a supernova explosion, its core could collapse to an even smaller radius, 12 to 20 kilometers, to form a neutron star whose matter density could be as high as one hundred trillion times greater than that of the Earth. On the assumption that subatomic particles more massive than a neutron were to populate the stellar core, particles generally called hyperons, the core's collapse could result in the formation of a hyperon star of even smaller radius.

---

*Although conventional astrophysics believes that bare stellar cores would be energetically dead and gradually cooling off, the physics of subquantum kinetics predicts instead that they would be continuously generating energy in the form of cosmic rays. For more about this astrophysics, consult my books *Subquantum Kinetics* and *Genesis of the Cosmos*.

In fact in 2002, the Chandra X-ray observatory detected one such star having a radius of only 6±2 kilometers. These too would make excellent cosmic ray sources for transmitting interstellar messages. The genic energy hypothesis predicts they would continue to radiate cosmic ray particles for an indefinitely long period after their initial formation, even after their supernova remnant had long since vanished.

Since we have done away with the lighthouse model and the notion of a rapidly spinning small radius star such as a neutron star, it is no longer necessary that the energy source powering a pulsar be exclusively a neutron star. A cosmic ray–radiating white dwarf or X-ray star would work just as well. In cases where no residual X-ray or optical emission is detected from the vicinity of a pulsar, we could conclude the source was either a bare core, a neutron star, or a hyperon star.

Now suppose a civilization possessed a technology that allowed it to beam electric fields to a distant location in a manner not unlike the way radio signals are transmitted: not only to beam them, but also to cause them to revolve axially in such a way as to generate lines of magnetic force. Furthermore, suppose that using this technology, this civilization had developed the means to generate such magnetic lines of force very close to the surface of the cosmic ray–radiating stellar core. These might be configured as a series of disks measuring about 0.5 to 50 meters in diameter and several meters thick and oriented parallel to the star's surface (fig. 44).

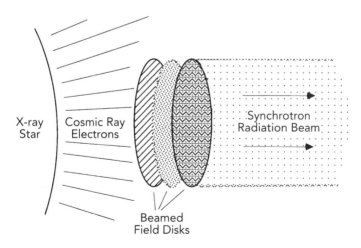

Figure 44. Field disks projected close to the surface of a neutron star for generating a stationary interstellar communication beam.

These fields need only have moderate strengths in the range of some tens of thousands of times the strength of the Earth's magnetic field. The incredibly strong magnetic fields that the lighthouse model uses in producing its synchrotron beams, fields 10^9 to 10^{15} times stronger than the Earth's field, are not mandatory in the ETI beacon model. Recall that lighthouse model theorists were forced to postulate enormous magnetic field intensities to secure a sufficiently strong spin-braking force so their model could account for the gradual lengthening of pulsar periods. There is no independent evidence that such strong fields exist, and as was discussed earlier, observational evidence instead suggests that such super-strong magnetic fields *should not exist*.

As the star's outgoing cosmic ray electrons move through the tele-metered magnetic field disk, they would emit synchrotron radiation and would beam this radiation in the direction of their outward flight. By deploying a succession of such disks of alternating magnetic polarity, the cosmic rays could be made to follow a straight-line flight path, and the radiation from these consecutive deflections could be made to add up to produce an intense beam of electromagnetic radiation capable of communicating over interstellar distances. Comparing this to the particle-beam-communicator design shown in figure 1, the stellar core would serve the same function as the particle accelerator module and the telemetered fields would serve the same function as the particle-beam collimator and modulator unit.*

By telemetering these field disks to the proper location on the star's surface, the synchrotron radiation beam could be aimed away from the star in any direction and targeted on a particular solar system hundreds or thousands of light-years away. Moreover, by properly modulating these fields, the beamed radiation could be made to produce precisely-timed integrated pulse profiles. The pulse period could be made to decrease at a precisely determined rate and to incorporate precise changes in timing and polarization so as to reproduce any of the pulse-ordering properties observed in pulsar signals. Mode switching would involve simply changing the kind of information being telemetered to the surface of the star. The transmitter producing these fields could

---

*Interestingly, some elements of this model, such as the idea that "standing wave" field patterns modulate the pulsar's cosmic ray particle flux, have appeared in some versions of the lighthouse model; see the discussion of PSR 0826–34 in appendix A.

be situated at a safe distance from the stellar core, perhaps positioned behind a distant planet or behind a force field shield where it might be sheltered from the outgoing cosmic ray wind.

A single stellar core or hyperon star could be used to simultaneously beam messages in many directions. Numerous field disks could be projected to various locations around the star's surface, each creating a beam of synchrotron radiation targeted at a specific part of the Galaxy. Thus, a given pulsar stellar source might serve as a navigation nexus, pointing beams to many different galactic locations. Also, each station might serve as a point for two-way communication, a kind of galactic Internet node, not only transmitting information but also receiving information from other locations using radio telescope apparatus. However, sending signals at superluminal speeds may require an entirely different type of communication apparatus.

Binary pulsars would present somewhat more of an engineering challenge. If the cosmic ray core star was orbiting a companion star, the synchrotron beam could be kept properly targeted on its destination by appropriately shifting the position of the telemetered fields so they tracked the star throughout its orbit. In the case of the EBM Pulsar, this orbit would be about four times the diameter of the Earth.

The power to maintain these projected fields could be derived from the stellar core's cosmic ray wind. The energy of this outgoing particle flux could be tapped through conventional magnetohydrodynamic techniques or through some more sophisticated means, perhaps where the projected beam itself serves as the energy-capturing net. Techniques to do this might be learned by pursuing research along the lines of the HAARP project, an acronym for High Frequency Active Aural Research Program.[16] This project, which was carried out in Gakona, Alaska, by the U.S. Department of Defense, used high-frequency radio waves to create a large plasma layer 50 to 300 miles up in the Earth's ionosphere and to induce resonances in that layer that would couple with natural oscillations excited by the solar wind.

Could a civilization possessing advanced field projection technologies perhaps even engineer an entire supernova explosion? By establishing a field bridge between the two poles of a star and inducing a resonant electrodynamic oscillation between these two linked regions, the star might be induced to explode. Clearly, a civilization must reach a high degree of maturity if it is to wield such a technology, for if it were to fall into the wrong hands, it could be used as a weapon of mass destruction.

While all this may sound like science fiction, a technology for projecting force fields to remote locations has been developed and tested by the military. More is said about this in the next chapter.

EIGHT

FORCE FIELD–BEAMING TECHNOLOGY

Aerial Plasmoids

As described in the previous chapter, it should be possible to produce decelerating forces on a neutron star's outward flux of cosmic ray electrons by artificially creating electrodynamic force fields near the star's surface, thereby producing a narrow beam of synchrotron radiation. Moreover, by properly controlling these fields, it should be possible to modulate the intensity of this beamed radiation continuum to produce pulses for interstellar communication or spacecraft navigation. If it was possible to telemeter force-inducing fields over interplanetary distances, say over several astronomical units (one A.U. being the distance from the Earth to the Sun), then such a feat could be possible.

Actually, the U.S. military is rumored to be testing a technology for application to warfare that could be doing precisely this—transmitting force-exerting fields to distant locations.[1] However, these methods would have to be deployed on a far larger scale if they were to be used for interstellar communication or for creating a superwave shield. Although details of the technology are enveloped in secrecy, its existence is evident from numerous sightings of strange luminous spheres in the night or evening sky. The devices that produce these effects are apparently able to ionize the atmosphere to such an extent as to form a luminous ball of plasma. These so-called *plasmoids* resemble in some respects the appearance of ball lightning, only they are far larger.

They are most probably generated by microwave beams. In his

article in *Nexus* magazine, Harry Mason quotes part of a December 1996 Voice of Russia interview with the Russian science authority Boris Belitsky. Belitsky said that in the 1993 Vancouver Summit, the United States and Russia proposed to cooperate in the testing of microwave weapons, and when the correspondent asked him how a microwave generator might be used as a weapon, Belitsky replied:

> It would be used to fire a plasmoid—that is, a blob of plasma—into the path of an incoming missile, its warhead, or an aircraft. The plasmoid would effectively ionize that region of space and, in this way, disturb the aerodynamics of the flight of the missile, warhead or aircraft and terminate its flight. This makes such a generator and its plasmoid a practically invulnerable weapon, providing protection against attack via space or the atmosphere.[2]

A sighting of one version of this novel technology occurred in the Superstition Mountains near the town of Apache Junction, which lies about 25 miles east of Phoenix, Arizona.[3] Elizabeth, her husband Dale, and their daughter Alicia had gone camping one weekday in the spring of 1993. Driving seven miles off the main road, they reached the campground parking lot at dusk. Leaving their car behind, they hiked up a canyon into the mountains. After journeying about a third of a mile, they set up camp. It was nearing eleven o'clock as they bedded down and began watching the stars in the beautiful desert night. They were the only ones out there. A wind was blowing, which thankfully was keeping the mosquitoes at bay. Alicia commented about how unusual this was, because she had gone camping in these same mountains a week before and had been eaten up by the mosquitoes.

As they were lying there, they were startled by a loud low rumbling that sounded like distant thunder. But the sky was cloudless. A few minutes later, looking down the canyon, they were amazed to see a huge aura of white light begin to fill the sky over the distant parking lot. It gradually brightened, looking like the kind of light that is seen filling the sky over a lit ballpark stadium. Some minutes later a fiery ball of white light rose from behind the mountain ridge on the far side of the parking lot and hung about 20 feet above the ridge, casting an eerie white light over the surrounding landscape. The sphere appeared to be about 2 degrees of arc in size (several full-moon diameters), which means that at that distance it would have measured about 60 feet in diameter. Being somewhat brighter than the full moon, it lit up the sagebrush and could be seen reflecting off their faces and clothing. It disappeared from view a couple of times but came back up. After five or ten minutes of this, the sphere then retreated behind the mountain, the aura faded away, and all became quiet again.

About 15 minutes later the rumbling began again and the aura of light reappeared. Within minutes a ball of light emerged again from behind the mountain. But this time, after clearing the ridge it moved to the left along the ridge and then veered out into the middle of the canyon hovering three to four hundred feet above the ground surface. They realized what they were witnessing could not have been produced by an ordinary searchlight. After some time, the sphere retraced its path, disappeared behind the ridge, and the aura once again faded away.

After another lapse, they again heard the rumbling sound. Again the sky lit up with a diffuse light. This time four luminous balls rose into the sky and lined up horizontally across the ridge. This horizontal formation moved to the middle of the canyon, where the spheres later moved to a vertical orientation. While still in full view, they suddenly disappeared. Then, in the blink of an eye, they reappeared in different positions. After some additional maneuvering, they again disappeared and the aura subsided. This sequence of events recurred several more times, the aura appearing, the multiple spheres emerging, moving in the sky, sometimes blinking out, the spheres retracting behind the ridge, and the aura subsiding. The entire display lasted slightly more than two hours. When the aura faded for the last time, the campers noticed that the wind also had ceased to blow and the mosquitoes had begun to invade their camp. To this day they have no explanation for the awesome sights they witnessed that night in the desert mountains.

Similar aerial electromagnetic technology displays have been seen in Australia. In a four-year period between May 1993 and May 1997 there were more than a thousand reports of aerial "fireballs" and associated luminous energy emissions that followed long trajectories over various parts of the continent. In his *Nexus* article, Harry Mason explains:

> These fireballs have been observed in all our Australian states . . .
> and in many cases have exhibited variations on and combinations of

the following actions: very-low-altitude, nap-of-the Earth trajectories; small-to-nonexistent tails; no fragment drop-off; apparent velocity often very slow and commonly less than that of sound; no associated sonic booms; considerable and sudden changes in course, as well as speeding up, stopping dead, reversing course and flying vertically upwards into space; creation of intense vibration of ground and housing during flypast; explosion in massive blue-white arcing light displays with major explosive sound events or silent, intense light-flashes; regular creation of power generation overvoltage outages and other electrical effects.[4]

He describes eyewitness reports of one event that took place on May 28, 1993. About an hour before midnight, a large, luminous plasmoid, or "fireball," was seen to fly from south to north in a low trajectory that was nearly parallel to the Earth's surface. It traveled about 1 to 2 kilometers at a subsonic speed similar to that of a 747 jet and emitted a pulsed, roaring sound as it went. Some described it as an orange sphere with a small bluish white conical tail, others said it was cylinder shaped and yellow-blue-white in color. Its low speed and near horizontal trajectory ruled out natural phenomena such as meteors as a possible cause. Upon impact with the ground, the fireball produced a near blinding, high-energy burst of blue-white light that rippled for three to five seconds. Witnesses said the illumination was as bright as day, allowing them to see for some 100 kilometers in all directions. The blast, heard up to 250 kilometers away, is estimated to have had an energy of more than two kilotons of TNT. A one-kilometer-diameter luminous hemisphere, deep-red-orange in color with a silvery outer lining, then arose and hovered above the explosion site, bobbing around for almost two hours before disappearing suddenly. The impact site was located by triangulation from seismograph records and from eye-witness accounts. However, careful examination of the site revealed no sign of a crater.

Mason reports another incident that occurred one night in mid-October 1994 in the Western Australia town of Tom Price. Residents reported seeing a red-orange fireball with a central "flame-engulfing black hole" cruising at about 100 miles per hour at an altitude of a few hundred meters. It was seen to be followed at evenly spaced intervals by two other fireballs tracking along the same trajectory. All appeared to originate from a military installation on the Exmouth Peninsula that was once a U.S. VLF radio communications base, but whose operation had later been turned over to the Australian government.

Microwave Phase Conjugation

The microwave generators that create these plasmoids may operate using the principles of *optical phase conjugation*.[5,6] The term *phase conjugation* refers to a special kind of "mirror" that is able to reverse the trajectories of the incident light waves and cause them to precisely retrace the path they followed to the phase-conjugating mirror. The outcome is as if the photons had been made to travel *backward* in time. If you shine a flashlight beam at an angle toward a regular silvered mirror, the beam will reflect off at an equal angle in the opposite direction. But if you angle a flashlight beam at a phase conjugate mirror, the returned beam will shine directly back at your flashlight!

Optical-phase conjugation is most commonly known for its use in military laser weapons systems for destroying enemy missiles. In this application, a laser beam is directed at a distant moving missile target and light rays scattered back from the target are allowed to enter the phase conjugator, a chamber containing a medium having nonlinear optical properties. In this nonlinear medium, the scattered rays interact with two opposed laser beams of a similar wavelength to form a hologram-like electrostatic light-refracting pattern called a grating. Once this grating pattern is formed, the system has essentially locked onto its target. A powerful laser weapon is then discharged into this holographic grating pattern, whereupon the coherent laser light reflects in such a way as to produce an intense outgoing laser beam that retraces the paths that had been followed by the incoming rays that had originally been scattered from the missile. Consequently, the outgoing laser pulse converges precisely back onto its missile target.

The microwave counterpart of a laser is called a maser. Just as a laser produces a coherent beam of light radiation, so a maser produces a coherent beam of microwave radiation. Consequently, it should be possible to phase conjugate microwaves by following techniques similar to those described in experiments with optical phase conjugation. Although most of this research is still highly classified, an increasing number of papers have begun to be published on this topic since the late 1990s. And in 1993 a basic patent on this technology (U.S. No. 5,223,838) was issued to Raymond Tang et al. of Hughes Aircraft Co.

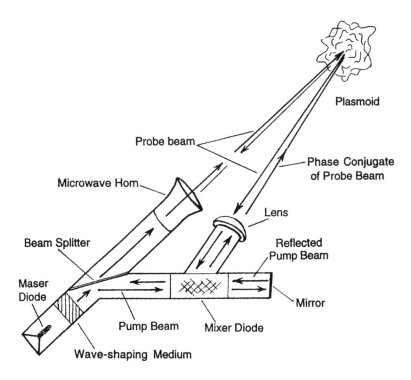

Figure 45. Proposed schematic of a phase-conjugate microwave resonator and associated aerial plasmoid.

applied to radar signal enhancement. Here we will give a rough idea of how this microwave technology might work in application to force-field beaming.

For example, suppose a bank of high-voltage capacitors is suddenly discharged to momentarily create a powerful air-ionizing arc, or irregularly shaped plasmoid. Suppose also that a beam of coherent microwave radiation from a maser is targeted on this plasmoid. This beam would be sent out by a phase-conjugating device something like that pictured in figure 45. This device splits the original maser beam into two beams, a *probe beam* and a *pump beam*. The probe beam is directed toward the plasma target, and some of its scattered microwaves shine back toward the phase-conjugator device and enter its mixer chamber. The *mixer* chamber is filled with a medium that has very nonlinear electromagnetic properties. Meanwhile, the pump beam that is split off from the maser generator is directed to the adjoining mixer chamber, where it passes through the nonlinear mixer medium, reflects from a wall at the far end, makes a second pass back through the mixer medium, and then reenters the maser generator compartment.

The microwaves that scatter from the plasmoid and enter the mixer medium strongly interact with the two pump maser beams to form an electrostatic grating pattern. This holographic pattern stores information about the directions and phases of all the scattered maser beam microwaves that have entered the mixer. The counterpropagating pump beams then reflect from this grating pattern to produce an outgoing microwave beam that *precisely traces the paths followed by the incoming scattered waves*. In this way, the outgoing microwaves converge back on the plasmoid, and from there travel back to the maser device from which they originated, the randomizing effects of any wave scattering having been automatically compensated for.

Physicists would call this return beam the phase-conjugated probe beam, or simply the phase-conjugate beam. The term *phase conjugate* signifies that the trajectories of the microwaves are identical to those that would be followed if the incoming probe beam microwaves had been made to travel backward in time. The grating that performs this time-reversed reflection is called a *phase-conjugate mirror* and the mixing arrangement that allows all this to happen is termed a *four-wave mixer*. Although the maser transmitter and four-wave mixer cavity are shown in figure 45 as being separated from each other, with proper engineering it should be possible to combine them into a common maser transmitter/receiver apparatus.

After the phase-conjugate beam returns to its maser source, it reflects back from the maser's end mirror, travels out once again as a coherent beam, scatters off from the plasmoid, and again returns to the mixer chamber. This closed-loop path from the maser beam generator, to the plasmoid, to the mixer, to the plasmoid, and back to the maser beam generator causes the maser system to function as a *phase-conjugate resonator* and to *preferentially select and amplify only those microwaves that target the phase-conjugating mixer chamber*. As a result, most of the microwave power emitted from the maser will end up being confined to a beam that extends between the maser beam generator, the plasmoid target, and the mixer chamber. The energy bottled up in this beam will then progressively build up to a very high value.

Thus, unlike a conventional maser, which emits a nonreturning energy pulse that eventually scatters into space, a phase-conjugating maser preserves the majority of its energy in an enduring beam that extends between the maser and its target. A portion of the maser beam energy, though, will go into heating the plasmoid. Also, to a lesser extent, some energy will be lost due to inefficiencies in the maser mixer cavity. The phase conjugator will tend to amplify those microwave trajectories that are most efficiently scattered back from the plasmoid. There will be a tendency, then, for the beam's energy loss to the plasmoid to be minimized. However, if the plasmoid were to intersect a solid target, the enormous energy stored up in the beam would suddenly be absorbed by the target, resulting in the target's destruction.

Tom Bearden, who has written several books and papers about this microwave technology, believes many of the unusual luminous aerial phenomena that have been observed are produced by military tests of phase-conjugating masers.[7,8] He points out that the incoming probe beam and its returned phase-conjugate beam would precisely superimpose with one another so as to mutually cancel their transverse force vectors. The two beams in combination would consist of *energy potential waves* (e.g., electrostatic charge density waves). Since energy potentials have magnitude but no direction, physicists term such waves *scalar waves* to distinguish them from Hertzian electromagnetic waves, which have a transverse vector polarity. Consequently, the phase-conjugating maser would produce a stationary *scalar wave* beam between itself and the plasmoid that would remain largely invisible to normal means of detection.

Bearden further maintains that if two such scalar wave beams are made to intersect, the interlocked probe and phase conjugate beams would decouple in this intersection zone, allowing normal electromagnetic waves. Plasmas have nonlinear electromagnetic properties. Hence, if two such scalar beams were made to intersect a plasmoid, the nonlinear plasma environment might cause them to interact with each other sufficiently to "unmask" part of their energy. By choosing slightly different frequencies for the two maser beams and modulating their amplitudes in the proper fashion, it should be possible to conduct interferometry and to thereby control the shape, size, intensity, and location of the plasmoid.

Tesla Waves

A phase-conjugate maser may be able to transmit energy to a distant target or extract energy from that target through energy exchange with the target. However, to exert mechanical forces on a distant target, or for that matter to exert forces on distant cosmic rays, it is necessary to use a microwave source that radiates pulses having a sawtooth-shaped shockfront, rather than a sinusoidal shockfront typical of conventional electromagnetic waves. The force-exerting ability of electrodynamic shocks was first discovered by Nikola Tesla in the late nineteenth century. Tesla designed multimillion-volt radio frequency generators that created these wave discharges, which began with an abrupt rise in voltage lasting a few nanoseconds, followed by a gradual voltage relaxation that transpired over a period of 1 to 10 microseconds. The pulses repeated with a frequency of about 0.1 to 1 megahertz and were conducted upward to a crowning spherical terminal from where they radiated into space. These "Tesla waves" differed from conventional Hertzian electromagnetic waves in that their field potentials consisted of purely longitudinal energy potential gradients with no transverse component. Tesla viewed them as alternate compressions and rarefactions that move longitudinally through a gaslike ether medium in much the same way that sound waves move through air. As discussed in chapter 3, conventional Hertzian waves and Tesla waves both find adequate description in the methodology of subquantum kinetics.[9]

Tesla found that when he stood near these shock discharges he could feel them as a great force or sharp pressure striking the whole front of this body.[10] These effects were most apparent as a stinging of the face or hands, which persisted even when he situated himself behind glass and metal shields as far as 50 feet from the shock source. He was able to collimate these wave discharges by energizing a terminal placed at one end of an open-ended tube. The tube emitted a ray beam that was able to push objects, charge objects, or burn holes through objects, depending on their composition. For some of his experiments, Tesla fabricated these ray tubes as heavy-walled glass vacuum tubes. On some occasions, they experienced such a high outward pressure when energized that they would explode in spite of their high vacuum.[11] Tesla ascribed these forces to the action of ether currents propelled forward by the ether shocks he was producing, although it is possible that the shocks themselves were the cause, a forward longitudinal force being more efficiently applied by the initial shock front and less efficiently counteracted by the more gradual relaxation gradient that followed.

Tesla found that his shock beams did not diverge like normal electromagnetic emissions, but remained collimated and coherent over great distances much like modern maser beams. If so, then perhaps optical phase conjugation might be achieved just as easily with a Tesla shock-wave generator as with conventional lasers or masers, which produce beams of sinusoidal Hertzian waves. Perhaps phase conjugation could explain how his Colorado Springs tower was able to transmit such an enormous amount of power to a receiver station located 26 miles away. Despite its great distance, the receiver tapped about 10 kilowatts of high-frequency power, enough to illuminate 200 50-watt bulbs to bright incandescence.[12] If his transmitted power had dropped off according to the inverse square of distance, as it does for normal radio broadcast, his receiver should have picked up microwatts, not kilowatts. The matter is more readily explained if Tesla's receiver had established a beamlike link with the transmitting tower through the phenomenon of phase-conjugate resonance.

Solid-state crystal devices called IMPATT diodes have been developed that produce sawtoothlike shock emissions, much like those that Tesla was producing, and pulse at microwave frequencies rather than at radio frequencies. Although IMPATT diodes on the open market operate at only several hundreds of volts, diodes rated for much higher voltages and power outputs quite likely are produced for restricted military use. By using such a solid-state device in place of a maser, it should be possible to produce a microwave-phase conjugator that generates Tesla shock-wave beams rather than Hertzian maser beams, in which case it would be possible to produce phase-conjugate beams capable of transmitting force fields to distant locations.

Deployed on a much larger scale with properly controlled interferometry, such a beam technology could be used to establish electrostatic field gradients near the surface of a stellar core or neutron star to deflect outgoing cosmic ray electrons and thereby produce a synchrotron communication beam. Pertaining to the ETI message, a similar technology could be used to set up a cosmic ray–deflecting force field shield capable of protecting a planet or even an entire solar system from an advancing superwave.

The Crop Circle Phenomenon

Other evidence of the existence of a field projection technology similar to that used in producing plasma fireballs may be found in the phenomenon of crop circles. Crop circles are large, highly organized patterns, usually appearing in fields of grain, such as grass, wheat, barley, or corn, where the crops have become mysteriously flattened to the ground in a very short time (see fig. 46). Often the matted areas have the form of circles or rings, but they can also appear as rectangles, linear swaths, and arclike shapes. Based on the various characteristics that have been reported, crop circle researchers have concluded that the circles are formed by some kind of advanced microwave beam technology.

Most crop circles appear overnight, although some have been reported to form in full daylight. One formation appeared in a field in between successive aerial surveillance flights spaced just half an hour apart. The crop circle researcher Colin Andrews reports that of a group of seventy people who claimed to have witnessed crop circles in the process of being formed, all agreed that the laying of the crop occurred in no more than 15 seconds! They described the event as

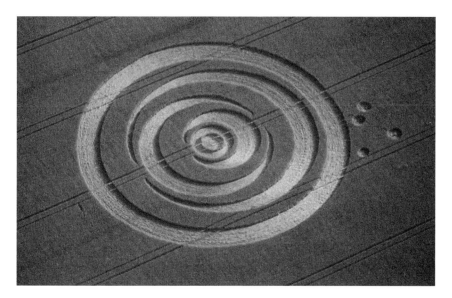

Figure 46. A 100-foot-diameter crop circle that appeared in a field at East Meon, Hampshire, England, in July 1995 (photo by Ron Russell).

being preceded by a sudden silencing of the songbirds, the stillness in the air being broken by a "trilling sound" accompanied by the sound of wheat heads banging together, despite an absence of wind. In one case the sound was captured on magnetic tape, and subsequent analysis found it to be beating at a frequency harmonic of 5.2 kilohertz.

Crop circles have been observed since the 1970s, with the documented sightings now numbering more than 10,000. Although most of the occurrences have taken place in the British Isles, formations have been reported in the United States, Canada, Australia, New Zealand, Tasmania, France, Switzerland, Russia, and Brazil. In earlier years the designs were relatively simple, consisting mainly of circular matted regions. Later, more elaborate designs began to appear, such as circles surrounded by a series of concentric rings. Often several circles might be found clustered to form well-defined geometrical patterns, sometimes interconnected by perfectly straight "avenues."

Beginning in 1990, the crop circle phenomenon dramatically intensified. Not only was there an exponential increase in their numbers, but there also was a dramatic increase in the size of the patterns. Every year heralded the appearance of new types of designs. Whereas earlier designs consisted mostly of simple circle and ring patterns ranging up to about 20 meters in diameter, now patterns began appearing that stretched out over several hundred meters. In addition, these new designs were far more complex.

Although disposed in two dimensions, some of the new patterns represented higher dimensional fractal forms like the "Mandelbrot set" crop glyph. The Mandelbrot set is generated by performing a specific mathematical operation on a complex number and repeating the operation by taking the product of the previous calculation and using it as input for the next calculation. Two of these complex patterns appeared in wheat fields in 1991, one in Ecleton, England, and the other at Barbury Castle, Wiltshire, England. Equally complex crop circle fractals depicting the Julia set appeared in July 1996. A single Julia set spiral composed of 151 circles stretching out over a diameter of 400 feet appeared in a wheat field near the ancient monument of Stonehenge. That same month a triple Julia set spiral appeared in a wheat field at Windmill Hill, England. It was composed of 194 circles that stretched out over a distance of 700 feet.

Within a crop circle's compacted region, the plant stems are found compressed to the ground so firmly as to leave their indentations in the soil, yet their stems are not broken or damaged. They sometimes splay radially outward from the circles' centers, but more often are angled so as to form Archimedian spirals, sometimes proceeding clockwise, other times counterclockwise, and on occasion a combination of both. The combined contrarotating spirals are particularly perplexing because the direction of the rotary lay abruptly flips at a well-defined boundary. On one side of this border, a stem would lie in one direction while its immediate neighbor on the other side of the border would lie in the opposite direction.

The crop circle that appeared at Cheesefoot Head in 1987 exhibited a clockwise spiral pattern that was deformed in the circle's northern sector to form a radial swath directed northward (see fig. 47). Near the center of the circle, this north-aligned pattern made a surprisingly sharp 90° flip transition to the spiral pattern. Yet another spiral crop formation had a central patch measuring two meters square in which the stems all faced in the same direction with the spiral pattern emerging outward from the square's periphery.

Braided or plaited lay patterns have also been found in which bundles of stems have been left overlaying one another in alternating directions, giving the appearance of a basket weave when viewed from afar. Crop circle researchers have puzzled over how these complex weaves could be formed. In their book *Circular Evidence*, researchers Pat Delgado and Colin Andrews write:

> Some of the bundles have had two or more bundles laid at differing angles over and under them so that they are actually intertwined. The force field that produced this would have to be operating like a knitting machine. On two occasions, in two different circles, some bundles of these braids have been laid in opposition to each other. There seems to be no limit to the complication of lay that this extraordinary unknown force can produce.[13]

Even more detailed crop knitting has been noted. For example, the crop circle researcher Ilyes observes that the crop within a circle is laid in bundles composed of anywhere from a few plant stalks to as many as forty or fifty.[14] The stalks in a bundle are aligned parallel to one another and are bound together by their leaves, which wrap around in crisscrossing fashion. Ilyes notes that, outside the crop circle

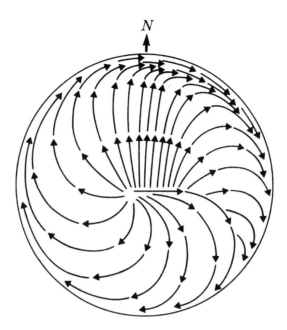

Figure 47. Schematic of a crop circle that appeared at Cheesefoot Head, England, in 1987 (Delgado and Andrews, Circular Evidence, *128).*

perimeter, the seed heads of the standing crop are randomly oriented. On the other hand, within each bundle, the seed heads are parallel to one another, as if commonly aligned by an applied force.

More than just mechanical force is involved in creating the crop formations. There is overwhelming and consistent evidence that the crop has been heated during the crop circle–forming process, for expulsion cavities called "blown nodes" are seen at the sides of individual stalks. The biophysicist Dr. W. C. Levengood produced similar results in the laboratory by inducing a very rapid rate of heating.[15] Ilyes postulates that crop circles are formed by a microwave maser beam projected down from above and that associated cellular heating effects cause the fluid within the plant cells to expand and erupt through a node or joint in the stalk's surface. She also suggests that the crop stalks suddenly wilt and become supple as their plant cells dehydrate and expand during this brief heating phase. In this way they are susceptible to being reoriented by the applied force without being broken.

Other tests carried out by Levengood have shown that the molecular and cellular structure of the plants has become substantially changed with distinct signs that they have been exposed to transient high temperatures.[16] Seeds taken from inside a crop circle formation have been found to be altered with respect to both their growth rate and their mode of germination. When the seeds are planted, their growth rates in some cases are seen to have become increased by as much as 45 percent. Moreover, extended observation of the crop circles shows that once the plants have been laid flat by this force, their head ends never again attempt to grow vertically.

Researchers have found that crop circles are pervaded by a residual energy field that remains for some weeks after the circle's formation. Investigators Ron Russell and Dr. Simeon Hein have used a portable TREK-520 electrometer to make electric field measurements of several crop circles in the Wiltshire region.[17] Their findings indicated the presence of a "membrane" at the outer edge of crop circles, across which the measured field energy shifts first lower, then perceptibly higher. These spikelike energy variations were also found in the interiors of crop circle, at the edges of internal details, and especially at the centers of the circles. They found they had to wait several days before entering a newly formed crop circle because otherwise the field energy would cause their electronic equipment to malfunction and their power-supply batteries to rapidly drain. It is commonly reported that the batteries of cell phones, GPS devices, magnetometers, video cameras, and tape recorders drain rapidly within crop circle formations. This calls to mind reports of car engines stalling and radios going dead during UFO close encounters.

The unusual growth phenomena, residual fields, and complex way in which the compaction patterns are formed have led researchers to conclude that crop circles cannot have been produced by common hoaxers using simple hand tools. Delgado and Andrews summarize the crop circle phenomenon as follows:

> Any specification for a force that can produce all these complications will have to include the following features: flattening crop stems of various thicknesses hard to the ground without damage, spiral whorl rotation, contra-rotation, multilayering containing multi-direction lays, blast-effect radial swathes, and all the different central area formations. Not only must the force be able to lay the crop in either direction of rotation, it must be able to do both rotations without gradation on the floor of the same circle.

It must be a strong force of short duration that induces horizontal growing into the plant, replacing its natural tendency to the vertical . . . It must also be able to make the root end of thick-stemmed plants pliable enough to bend to a sharp, almost 90° angle, without fracturing and disturbing the plants' growth rate.

It must be able to construct a flattened ring around the outside of a circle, closely following the contour of the circle wall. Besides creating circular shapes, it must also be able to flatten a dead straight pathway several meters long. It should have the ability to miss narrow, arc-shaped areas of crop and so leave these stems standing, like a low, slim, curved screen. It must be so violent that some plants are pulled up randomly or ejected from the soil and thrown into the peripheral standing crop. It must be quiet.

It is very sobering to stand in one of these circles and ponder what force could have arrived and departed, leaving behind this beautiful record of its visit with no clue as to how it was achieved.[18]

One farmer who witnessed the creation of a ringed circle in his small corn patch in Romania described the force as so strong that it tore away his hat and flung him to the ground. It was accompanied by a "terrible whistling sound."[19] This same force could explain why a jackrabbit carcass was found inside one crop circle dehydrated and *compressed to a pancake*. The crop circle researcher Donna Higbee describes several cases in Canada and England in which crop circles have been found to contain compressed porcupines. Porcupines, which normally stand at least 12 inches tall, have been compressed down to *two to three inches* by some tremendous force. She notes that porcupines stay put when they detect danger, relying on their sharp quills for defense. Hence, there is less likelihood that they would flee at the time the circle was being formed. Higbee also reports instances in which porcupines were forcefully dragged into a crop circle. She writes:

Also, there have been at least two cases where a skid mark in the soil, embedded with broken quills, showed the unfortunate porcupine had been dragged by force from the edge of the formation into the center. Its quills were oriented in the same direction as the swirled wheat.[20]

A technology capable of projecting a force field to a remote farm field to bend stalks of wheat or blades of grass could also be made to project a force field that could bend the trajectories of cosmic ray particles in space. Moreover, a force field technology able to create intricate patterns in fields of vegetation could also be made to modulate forces near the surface of a cosmic ray–emitting star to produce synchrotron pulses similar to those observed coming from pulsars. Are the hundreds of crop circles being produced each year the result of force field projection experiments being carried out by our own scientists working in secret military programs? Or are these messages being left by extraterrestrial visitors who are trying to display for us the same technology they use to produce pulsar signals and star shields?

An ETI Connection?

There have been many cases where unconventional flying objects have been sighted either at the time crop circles were being formed or just prior to their appearance. In none of the documented cases has anyone seen or met the occupants of these vehicles, so we cannot be certain whether they are from here or from out there. Nevertheless, given that these vehicles utilize a very advanced method of propulsion that could be used for flight beyond the Earth's atmosphere, the idea that they may be piloted by extraterrestrials remains a very real possibility.

Delgado and Andrews recount one incident that took place in England over Silbury Hill.[21] A woman was driving home late one night in July 1988 when she noticed a large, golden, disk-shaped object hovering within the cloud cover. A bright parallel beam of white light projected downward from the disk at roughly a sixty-five-degree angle to a spot a few kilometers away. While she was observing this, a "surge of energy" passed through her car, at which point half a dozen articles that had been stored in a pocket on her dashboard suddenly rose up and flew backwards onto her lap and onto the passenger seat next to her. One day later, ten crop circles were discovered at about the location where the beam had been directed. Another five circles appeared several days later, and still others were discovered two months later.

Another sighting was made in 1991 by two Japanese boys. Roy Dutton, who recounts their story, writes:

> As the boys watched, a pillar of "transparent white steam or smoke" [was] generated downwards from beneath the hovering [glowing orange object]. The pillar revolved and grew wider towards the base, "giving the impression of a trumpet." When the base came into contact with the grass, a flattened ring was produced, about 30 cm wide . . . Immediately after the [grass was flattened], the "trumpet" was retracted, and the object shot back into the sky. . . . One of the young men commented that he felt "a warm wind and drops of water on [his] face" as he watched the grass being whirled down. As he stood transfixed, he reported that he simultaneously heard a "low alternating sound" transliterated as "gu-on, gu-on."[22]

The crop circle photographer Steven Alexander has shot a video that shows a luminous sphere meandering over a crop circle days after the circle had been formed. One eyewitness who saw the sphere up close said it was about the size of a basketball. Another researcher, Peter Glastonbury, witnessed luminous spheres in a crop field near his home in England. He states:

> After fifteen minutes, we heard sounds that were very much like the crackling you hear around your head when taking off a heavy woolen sweater. My daughter and I both saw a small ball of light hovering in the crop. Through binoculars, I could see several small balls orbiting into a central point. It was dipping in and out of the crop. The next day we discovered a formation exactly where we had heard the crackling sounds and saw the light the night before.[23]

UFO landings in fields have also been found to leave behind circular swirl patterns. These are presumably produced by a saucer's propulsion beam. In Australia, these lay patterns have become called UFO "nests." These usually have a much rougher form than crop circle designs, their edges not being as sharply defined. The vegetation within them exhibits signs of nonchemical withering, and analysis has shown that some of the changes could have been produced by powerful microwave radiation.[24]

The NASA scientist Paul Hill has gathered considerable information from numerous UFO sightings suggesting that force field beams are commonly used by UFOs as a method of propulsion.[25] One of the documented UFO encounter cases he reviews took place in Greece near the villages of Digeliotica and Agiou Apostolou.[26,27] One evening in February 1959 people living there reported seeing a luminous disk emitting a humming noise and circling over the two villages for about 10 minutes. Radios failed to operate and the current in one house went dead. The disk flew low over one house and, as it did, the force of its propulsion beam caused the house to shake and its roof tiles to clatter loudly, making its occupant think there was an earthquake. When the villagers later inspected the house, they found many of its roof tiles had been displaced and that others were lying on the ground.

We might theorize that in producing its propulsion beam, a UFO spacecraft is projecting downward an intense column of Tesla-type microwave radiation—that is, shock-type microwave emissions that have the ability to exert forces on the material they intercept. As Tesla observed, shock emissions tend to collimate themselves into a beam. However, to further ensure that the beam's energy is not scattered and dissipated once it strikes the ground, the UFO propulsion system might utilize the technology of phase conjugation. Accordingly, the microwaves reflected back from the ground toward the spacecraft propulsion apparatus would enter a mixer chamber that would generate an outgoing *phase-conjugate* beam of the ground-reflected "probe" beam, the two counterpropagating beams being locked together in phase with one another. Consequently, the downward-directed phase-conjugate microwaves the spacecraft was emitting, upon reflection, would faithfully return to the spacecraft propulsion apparatus. The spacecraft propulsion apparatus and the Earth's surface, then, would together function as a phase-conjugate resonator, allowing microwave radiation to build up within the beam to a very high intensity. The upper end of this resonant column of "solid" radiation would push up against the spacecraft as the lower end pushed down against the Earth's surface, thereby lofting the spacecraft against the Earth's gravitational pull.*

---

*Research carried out by James Woodward, a professor of physics at Cal State Fullerton, indicates that electromagnetic waves can induce lofting forces in piezoelectric ceramic media. His ideas are described in a 1994 U.S. patent (No. 5,280,864) and in a physics journal article (*Foundations of Physics Letters*, vol. 3(5), 1990). Woodward has conducted experiments that confirm this thrust effect in the audio frequency range (~10,000 Hertz), and his calculations suggest that this thrust may be substantially increased at higher frequencies, with optimal performance being obtained in the microwave range (0.1 to 10 gigahertz).

The microwave beam technology that a UFO uses to buoy itself upward may be the same one being used to form crop circles. UFO sightings indicate that these craft can project several beams at one time and can control not only the direction in which they aim the beams but also the degree to which the beams diverge. By studying the intricate crop circle patterns, we can learn something about the sophistication of this propulsion-beam technology. The beams would need to have a resolution of the order of just a few millimeters to create crop lay patterns having the kind of sharp-edged boundaries that are observed. Also, they would have to be capable not only of pushing but also of pulling, in some cases even uprooting individual stalks. To produce the swirling of the lay patterns, the beam-phase conjugator apparatus must be able to induce a vortical component to its beamed force field. The unusually complex weaving and braiding seen in the crop circle patterns suggests the creators of these designs have very precise and rapid control over their microwave beam "paintbrushes."

The 5,200 Hz ultrasonic beat frequency that has been detected in one crop circle and the *gu-on, gu-on* sound that was reported to be heard at another sighting would seem to suggest the crop circle makers produce beams that contain many microwave frequencies whose difference frequencies produce harmonics in the sonic or ultrasonic range. Moreover, by properly phasing these various frequencies relative to one another, the beam generator could be made to control the waveform shape of these summed waves and hence the intensity of the force they generate. By using two or more beam generators, it should be possible to employ conventional microwave interferometry techniques to control the intensity and direction of these forces across the diameter of a crop circle pattern.

I heard a story from a friend that may be relevant here. In the early '90s she had attended the annual conference of the Society for Scientific Exploration, which was held at Princeton University. Mr. Jean-Jacques Velasco, head of GEPAN/SEPRA, the French space agency program involved in investigating UFOs, was at the time lecturing and was being aided by the French UFO researcher Jacques Vallee, who had introduced him and helped as a translator. At the end of his lecture, a member of the audience asked Velasco about the crop circle phenomenon. Velasco began to respond in French, saying, "Yes, that is very secret, they are produced by beams from the orbiting Mylar microwave platform." At this point, Vallee cut him off, whispering in French, "*Tessez-vous!*" (Shut up!). Velasco then changed the subject. Apparently, Velasco considered the crop circle phenomenon to be man-made and part of the testing of a new kind of microwave weapon technology. An orbiting space platform made out of Mylar plastic would have been invisible to radar detection.

The same force field projection technology that is used for UFO propulsion, and which also appears to be used in highly secret military weapons projects for producing aerial fireballs and possibly crop circles, may be the same technology that extraterrestrial civilizations are using to produce pulsar signals.

Setting Up a Star Shield

If such a force field technology were possible and could be used for interstellar communication, it might also be used as a means of deflecting superwave cosmic rays so that they would pass harmlessly around a star and its planets. For example, it might be possible, through such field telemetry, to produce a force field "shield" in the direction of the Galactic center to shelter our solar system from an approaching superwave. Deflected from their flight paths by just half a degree, approaching cosmic rays could be made to miss our solar system and thereby could be prevented from propelling cosmic dust into our interplanetary environment. The shield, however, would need to be huge—over a tenth of a light-year in diameter—and would need to be positioned sufficiently far from our solar system, again about one tenth of a light-year. In addition, creating it would require an enormous amount of total energy. But because the shield would function as an energy-storing phase-conjugate resonator, the shield's fields could be built up to the required intensity by using a moderate amount of supplied power.

Could the tubular jet seen to project from the northern side of the Crab supernova remnant (figure 39) be an example of such a shield being deployed on a much larger scale? In this case the fields would be configured as an immense cylindrical shell measuring one and a half light-years in diameter and two and a half light-years or more in length, with magnetic forces oriented along the length of the cylinder and concentrated at its outer perimeter. Cosmic ray electrons passing through this cylindrical shell would become either captured in spiral orbits within the shell or deflected around the shell. The question remains as to whether it is possible to develop phase conjugate structures this large from a "point source" transmitter considering the required roundtrip time for an outgoing and return pulse. Would it be possible to produce a phase-conjugate shield this large even using superluminal shocks?

Many have wondered what the reason is for the continued appearance of crop circles—what have the crop circle makers been trying to communicate to us all these years? If crop circles are being made by visitors from nearby star systems who are members of the galactic communication network, perhaps they are continuing to create these formations as a way of giving us blatant demonstrations of the technology we hopefully will one day develop ourselves to protect our planet. By means of the pulsar beacons, the galactic community has been demonstrating this technology from afar. Now, with crop circles it appears we are being shown the use of this technology virtually under our noses. Governments developing microwave phase-conjugate technology for military warfare could redirect their efforts toward setting up a force field shield for the purposes of planetary defense.

Contact

There were several reported instances during the twentieth century in which unusual radio or television transmissions had been received from space, some occurring in connection with aerial sightings of unconventional disk-shaped flying objects.[28] However, because they were transient events, they could not be later checked and verified, leaving disbelievers unconvinced that signals may have been received from extraterrestrial intelligences. The crop circle phenomenon has the advantage that the "communication," the crop matting patterns, stay around for up to several weeks, long enough to allow many investigators to photograph and study them. Such investigations have developed into a scientific discipline that has become known as *cereology*. Still, there are die-hard skeptics who refuse to believe people's reports of crop circles forming in a matter of seconds and who have convinced themselves that the hundreds of these designs produced every year in many countries around the world are all the work of a few well-financed pranksters.

Like crop circles, pulsars are a persistent phenomenon, although they endure over a much longer period, centuries or millennia rather than just weeks. As a result, pulsar signals have been diligently studied by many astronomers and the received data have been published in peer-reviewed scientific journals, examined, and reexamined. Thus, there is no question that these sources exist. The issue is their *interpretation,* and as history has shown, paradigm shifts do not happen over night. Nevertheless, evidence of an ETI origin for pulsars may be easily checked out by consulting the published data, whose existence and accuracy are well accepted.

Perhaps we have just scratched the surface in our attempt to decipher the messages that pulsars convey. Buried within the rapid shifts in the intensity and polarization direction of their signals may be a wealth of information still waiting to be decoded. We should launch a major program that would employ expert cryptographers to assess the information content of these transmissions. Perhaps today's SETI researchers might join the effort. The Open SETI program proposed by Gerry Zeitlin is a good start.[29]

The team of researchers could include remote viewers, people who are able to use their mental powers to "see" events taking place at a far-off distance. The viewers might be given the coordinates of certain pulsars and asked to report what information if any they sense coming from those locations. Although there would be no way of directly verifying the validity of what the viewers would be reporting, their perceptions might provide useful guidance to help SETI astronomers locate and later decode intelligible transmissions.

We should not expect all pulsar beacons to be serving as information conduits if, in fact, their main purpose is for spacecraft navigation. But the Vela pulsar should be a good candidate to begin with. It is the brightest radio pulsar in the sky and has a high signal-to-noise ratio. Also, it appears to be a symbolic indicator of the 14,100 years that have elapsed since the last major superwave passed through the solar system (chapter 6). The Crab pulsar may be another candidate for study. Being carefully placed on the forefront of the 12,150-B.C.E. superwave event horizon and at the center of a scale-model map of this event horizon, we might surmise that the subject of its transmission has something to do with this superwave event. Its pulse-to-pulse radio intensity variations are among the largest known for any pulsar.[30] Other good candidates to study are long-period pulsars having complex time-averaged pulse profiles. The pulse-to-pulse signal variability has been found to be significantly higher in such pulsars, suggesting the possibility that their signal modulation may contain transmitted information.

We may already know enough about the Galactic core explosion message of the pulsar network to allow us to devise a return message that would let nearby galactic civilizations know we are aware of their transmissions. We could re-create the Crab–Vela pulsar arrow on the ground by situating a pair of Tesla shock wave transmitters at some distance from each other, one pulsing at the rate of the Crab pulsar and the other pulsing at the rate of the Vela pulsar. Nearby extraterrestrial visitors who were members of this galactic collective might find such a transmission of particular interest. The transmissions would be symbolically laying out an arrow along the ground, inviting a close encounter or landing somewhere beyond the tip of the "Vela pulsar" transmitter. In effect, we would be creating a contact scenario similar to that portrayed in the movie *Close Encounters of the Third Kind*. But the motive for doing this would go beyond just the thrill of communicating with visitors from a more advanced civilization. As novices being indoctrinated into the galactic "federation," our top priority should be to ask them for their knowledge and help to assist our civilization survive the next superwave. We already have been given the invitation to do so.

APPENDIX A

ORDERED COMPLEXITY

The following intends to show not only that pulsar signals have a high degree of ordering, but also that there are many types and levels of ordering in their signals. In this way, a case may be built for arguing that pulsars are most likely of artificial origin. The premise here is that highly ordered signal characteristics or the existence of precise cyclical periods would indicate that the source had a high likelihood of being of artificial origin. Such characteristics contrast with most natural astronomical phenomena, which as a rule exhibit random modulation of their signals (signal noise) or, if they exhibit cyclical variations, their cycles typically have limited precision. Simply stating that the highly ordered signal characteristics of pulsars is a consequence of the very rare and highly compressed condition of matter found within neutron stars is insufficient reason to abandon consideration of the ETI alternative: the neutron star lighthouse model, with its inferred highly ordered emissions, has no independent observational verification. Astronomers have presumed its existence for the express purpose of explaining pulsars as a phenomenon of nature.

The following does not attempt to determine if an intelligible message is being carried in the signal of any one pulsar. It seeks only to establish the high degree of ordered complexity the signals exhibit and the difficulty that would be encountered in explaining them as arising from naturally occurring sources. Let us now review some of the ordering properties of pulsars.

Pulses and Time-Averaged Pulse Profiles

Pulsar astronomers use the term *subpulse* to refer to the individual radio beeps coming from a pulsar. We will instead continue to use the term *pulse,* which should be less confusing to the reader. Pulsar astronomers use this term because they consider the time-averaged profile to be the true pulse feature and the individual pulses forming it as being one step down in the informational hierarchy. This terminology is undoubtedly influenced by the lighthouse model, which interprets the precisely-timed time-averaged profile as the basic datum of neutron star rotation speed.

A given pulse typically lasts a few thousandths of a second. When its change of intensity, or *amplitude,* is plotted out as a function of time, the resulting trace usually has a hump shape, a simple rise and fall. Furthermore, successive pulses typically vary in both intensity and pulse cycle phase. This variation, however, is *nonrandom.* That is, when many pulses are summed up, they collectively produce a highly invariant pulse-intensity profile. We refer to this summed profile as the *time-averaged pulse profile.*

The time-averaged pulse profile constitutes a level of informational order situated one hierarchical step above the pulse level. It is built up from pulses in much the same way that a computer image is built up from the series of information bits that compose its image file. The pulses, the pulsar signal's subsidiary information bits, in combination produce an "image"—the time-averaged profile.

Examples of pulse sequences received from each of four pulsars are shown in figures A.1 and A.2. The horizontal time scales shown here are measured in terms of degrees of pulsation-cycle phase, rather than in seconds of time, where 360 degrees equals one complete cycle.* Each graph presents a series of 14 consecutive pulse traces. The variously sized ellipses superimposed on the pulse traces denote the intensity and direction of radio signal circular polarization. The open and filled ellipses, respectively, denote right-hand and left-hand circular polarization. When superimposed one upon the other, these pulse profiles build up the composite time-averaged pulse profile shown at the top of each graph, a contour that is unique for each pulsar.

---

*A pulse appearing at a phase of 90° would be displaced from the time reference point by one fourth of the 360° pulse cycle.

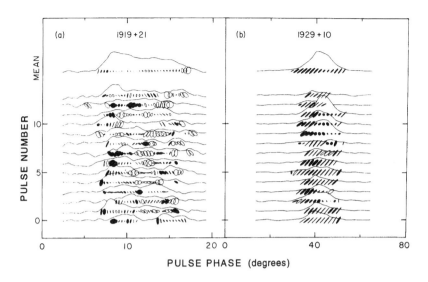

Figure A.1. Pulse sequences and time-averaged pulse profiles for: (a) pulsar PSR 1919+21 and (b) pulsar PSR 1929+10 (Manchester, Taylor, and Huguenin, Astrophysical Journal, *figure 14).*

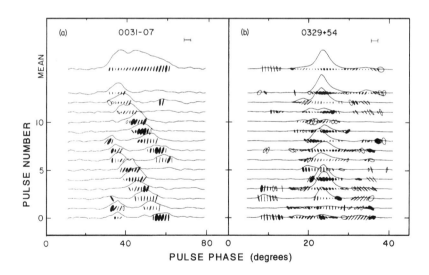

Figure A.2. Pulse sequences and time-averaged pulse profiles for: (a) pulsar PSR 0031–07 and (b) pulsar PSR 0329+54 (Manchester, Taylor, and Huguenin, Astrophysical Journal, *figure 1).*

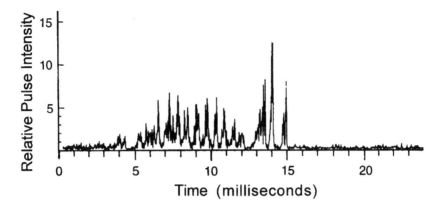

Figure A.3. A typical pulse profile for PSR 0950+08, showing rapid intensity variations (Hankins, Astrophysical Journal, *figure 5).*

Although a pulsar's signal intensity, polarization, and arrival time vary considerably from one pulse to the next, the shape of its time-averaged pulse profile remains constant. Moreover, the polarization structure within the envelope of the time-averaged pulse profile—the manner in which the time-averaged pulse polarization varies with pulse cycle phase angle—also remains constant. The phase position of the time-averaged pulse profile within the pulse cycle is also constant, as is the length of the pulse cycle.

More than half of the observed pulsars have been found to have pulses that are themselves composed of a still finer pulsation substructure that consists of a train of very rapid *micropulses.* Micropulses typically last a few hundred microseconds. In some cases they recur with a well-defined periodicity that is distinct from their pulsar's primary period. The pulses of PSR 2016+28, for example, are modulated with micropulsations having a recurrence period of about 900 microseconds. PSR 0950+08 is another example. A small fraction of its pulses contain micropulse trains with oscillatory periods ranging from 300 to 700 microseconds, over 1000 times shorter than the pulsar's primary period, which measures about a quarter of a second. An example of one modulated pulse received from this pulsar is shown in figure A.3.

Pulse Modulation

In some pulsars, the intensity of a given pulse is seen to correlate in a specific way with the intensity of pulses that preceded it. For example, there are pulsars in which the intensity of consecutive pulses oscillates in a regular fashion, signal strength being observed to alternately wax and wane over a series of pulses. In other pulsars this periodic *pulse modulation* is present only when every other pulse in a sequence is sampled. The period of such a modulation cycle, customarily designated as P_3, typically ranges from 2 to 20 times the pulsar's primary period. The primary period, which characterizes the average time between successive pulses, is designated as P_1. Compared with the primary period, the pulse modulation period is far less precise. Even in pulsars having a well-defined pulse modulation pattern, P_3 is precise to no more than two or three decimal places.

Further adding to this complexity, pulse modulation is often found to occur *only at certain phase angle positions in the time-averaged pulse profile*. Consider, for example, PSR 1919+21, the original pulsar discovered by Jocelyn Bell. The left-hand side of figure A.4 displays this pulsar's time-averaged pulse profile, also seen in figure A.1a. The profile has been rotated by 90 degrees so that its phase angle axis is plotted here vertically. Furthermore, the vertical phase angle axis has been segmented into 34 cycle phase "windows," and the pulse-modulation frequencies within each of these phase windows have been separately analyzed. The frequency distribution for each of these windows has then been plotted to form the series of traces to the right. A peak in any of these traces indicates the pulse-per-second frequency at which pulses recur at that particular phase in the pulse cycle. For example, the lower traces, which correspond to the leading part of the time-averaged profile, show peaks at around 0.21 and 0.24 Hertz, indicating that pulses wax and wane about every 4.2 and 4.8 seconds, or at about 3 times the pulsar's primary period. Farther up, the traces indicate no discernible modulation for the middle part of the pulse profile. Still farther up, traces corresponding to the trailing part of the pulse profile indicate again strong pulse modulation at similar frequencies, with a hint of a 0.44 Hertz modulation, second harmonic, appearing in just one or two of these traces. This frequency-versus-phase analysis forms a personality fingerprint of this pulsar's pulse profile that amazingly does not change over extended periods of observation. It is such ordering complexity that leads us to ask whether pulsar signals may be of artificial origin.

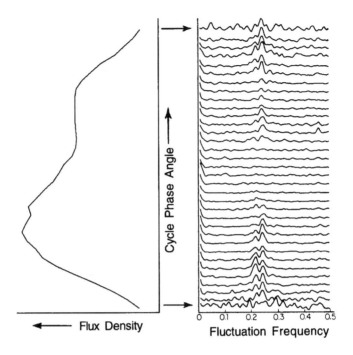

Figure A.4. Left: the time-averaged pulse profile of pulsar PSR 1919+21 showing intensity plotted against cycle phase. Right: fluctuation spectra observed in 34 time channels each corresponding to a phase segment of the time-averaged pulse profile (Backer, Astrophysical Journal, figure 1).

Pulse Drifting

One of the most unusual ordering phenomena observed in pulsars is that of pulse drifting. In *pulse drifting,* each successive pulse is slightly displaced in phase, giving the appearance that over time the pulse gradually moves across the time-averaged pulse profile. One example of this is shown in figure A.2a. Usually pulses drift from right to left—that is, from the trailing portion to the leading portion of the time-averaged pulse profile. However, in some pulsars the reverse drift direction is also observed. As a pulse approaches the leading edge of the profile, it weakens and vanishes within a few pulse periods. Meanwhile, a new pulse arises near the trailing edge of the profile to continue the drifting process. The pulses repeatedly scan across the time-averaged pulse profile with a characteristic period P_3, equal to the pulse modulation period described earlier. Moreover, in some pulse cycles, the leading and trailing pulses will appear together, separated by a characteristic time interval usually designated P_2. This pulse-spacing interval is yet another invariant feature observed in pulsars.

Each pulsar that exhibits pulse drifting has its own unique variety of drift activity. It is worthwhile describing here in some detail these variations to show that pulsar signals can have a remarkable variety of extremely complex ordering modes.

Linear Drifters. Some pulsars are classified as *linear drifters* because consecutive pulses drift across the time-averaged pulse profile at a constant rate. Thus, when successive pulse cycles are mapped out, consecutive pulses form a series of straight diagonal lines, like those seen figure in A.2a. In this particular pulsar, PSR 0031–07, successive sequences of drifting pulses maintain similar drift rates. However, there are other linear drifters in which each successive drift sequence maintains a drift rate different from the preceding sequence, with no particular pattern being apparent in the evolution from one drift rate to the next. When the pulse sequences from such a pulsar are mapped out, they form a series of diagonal strings, each having a different slope. PSR 2016+28, whose time-averaged pulse profile is composed of two components, exhibits the additional peculiarity that linear pulse drifting tends to occur more frequently in one component than in the other.

Nonlinear Drifters. In certain other pulsars, the pulses progressively change their drift rate as they scan across the time-averaged pulse profile. These types of pulsars have been classified as *nonlinear drifters*. For example, in PSR 1919+21, pulses drift rapidly into the pulse profile envelope from the profile's trailing edge, decelerate as they pass the center of the pulse profile, and accelerate as they drift out at the profile's leading edge. Thus, each pulse drift sequence executes an S-shaped trajectory when consecutive pulse cycles are plotted out next to one another.

Nonrandom Pulse Patterns. Finally, some pulsars have such complex pulse sequences that it is not possible to trace their pulse drift paths. Yet definite patterns of pulse recurrence are detectable. In PSR 1133+16, for example, brief intervals in the data sequence are discernible in which pulse patterns composed of three to six consecutive pulses occasionally recur. These patterns consist variously of single pulses that appear in either of the two pulse profile components, pulses that appear in both components, and null pulses—pulses that fail to occur.

PSR 0329+54 is another pulsar that exhibits nonrandomly patterned pulses. When a strong pulse appears in component-1 of its four-component pulse profile, this event is generally preceded in component-3 by several weak or missing pulses and followed by strong pulses. Also, when pulses appear in component-4, they are often accompanied by pulses in component-3 and followed one period later by a strong pulse in component-3.[1] The complexity of this organized behavior begins to approach that found in simple computer logic circuits.

Both pulse drifting and pulse sequence patterning are very unusual ordering phenomena that present a severe challenge to theoreticians seeking natural explanations for pulsars. Although possible natural causes have been suggested for this ordering in individual pulsars, astronomers are far from having a successful self-consistent theory that explains all types of pulse ordering. Such varied and complex signal ordering, however, might be expected if pulsars are beacons of ETI origin, their creators making extraordinary efforts to ensure that their signals are not misinterpreted as being of natural origin.

Multi-component Correlated Drifting. In most pulsars, the time-averaged pulse profile spans a small portion of the total pulse cycle. However, in pulsar PSR 0826–34, *the profile stretches over the entire pulsation cycle.* It is one of the few pulsars known to have such a broad pulse profile in the radio spectral region. Its broad 3-component profile allows it to accommodate five, or on occasion six, well-defined bands of linearly drifting pulses, *all of which drift across the time-averaged pulse profile in sync with one another.*[2] This pulsar is unique in that no other pulsar has so many bands of drifting pulses, spanning 45 percent of the entire pulse cycle. In no other pulsar does pulse drifting occur over such a large part of the pulse cycle.

In most pulsars that exhibit pulse drifting, pulses normally drift in one direction, usually from the trailing to the leading edge of the pulse profile. However, in PSR 0826–34, the pulses drift first in one direction, stop, and then drift back in the *reverse* direction, changing their direction at irregular intervals. All the while they stay in lockstep. This drift reversal phenomenon has been observed to date only in this pulsar. Another unusual feature is that the pulse drift rate progressively increases with each successive pulse cycle until the drift rate reverses, at which time it progressively slows down from its initially high drift rate.

Furthermore, as these pulses drift, they are kept spaced from one another by a fixed time interval, equal to 8 percent of the primary pulse period ($P_2 \sim 0.081 \, P_1$, or 0.15 second). In any given pulse cycle, five, or sometimes six, pulses will appear, each spaced from one another by this fixed P_2 interval, *regardless of how fast the pulses happen to be drifting.*

Yet another unusual feature of this pulsar is that its pulse transmissions experience long interruptions, called pulse nulling. These blank intervals have varied lengths, ranging from several seconds to more than seven hours. Although pulse nulling has been observed in a few other pulsars, PSR 0826–34 is unusual in that it stays in its nulling state for at least 70 percent of the time. As in other nulling pulsars, the phase positions of the pulses *freeze* during the nulling period, but this pulsar is unique in that the phase positions of all five or six bands of drifting pulses become frozen.

Observations of just this one pulsar shook the foundations of pulsar theory, sending all previous versions of the lighthouse model to the trash can. Previously, pulse drifting had been explained by supposing that individual pulses were produced by cosmic ray–particle cascades, or "sparks," that precessed around the star's magnetic pole as a result of the pole's magnetic field acting on electric fields in the cascade; the stronger the magnetic field, the greater the drift rate. However, such models predict that pulse drifting should proceed at a constant rate and in only one direction.[3] To instead account for the observed continually changing drift rates and sudden reversals of drift direction, the lighthouse model is left to the absurd prediction that the neutron star's magnetic field axis must progressively change its orientation relative to the star's spin axis and that at times it negotiates radical flips. Moreover, we are asked to believe that in engaging in these gymnastics, this powerful field, which is a trillion times stronger than the Earth's magnetic field, for some reason does not alter the even-spaced timing of these precessing particle cascades. Furthermore, the lighthouse model fails to explain why the pulsar's entire emission process mysteriously shuts down for periods of seven hours or more and somehow *remembers* the cycle phase at which its drifting stopped prior to the time its emission shut down!

In reporting this pulsar's unusual behavior, the pulsar astronomers J. Biggs, Peter McCulloch, P. Hamilton, Richard Manchester, and Andrew Lyne conclude:*

---

*The term "subpulse" used in the quote below signifies *pulse* according to the terminology that is adopted here in this book.

Current proposals for the drifting subpulse mechanism do not adequately describe the observed drifting subpulse behavior.[4]

A viable lighthouse model must be able to explain the variable drift rates with drift reversals, the correlation of the multiple drift bands, the constancy of the P_2 pulse spacing, the extended pulse nulling, the pulse "freezing" that remembers the pre-null pulse phase, and the observation that radio continuum is transmitted during the entire pulse cycle. But, such a model runs the risk of being unusually complicated and contrived. J. Biggs and his colleagues have sketched out a way in which the lighthouse model might be modified to account for a few of these unexplained properties—the correlated pulse drifting and drift reversal phenomena. They suggest the possibility that the neutron star's pulse emission intensity is *modulated by two standing wave field patterns that reside near the star's surface* and whose modulating effects interfere with one another so as to create the observed pulse phase-drifting behavior. They presume that these standing waves are somehow generated naturally and that they rotate along with the neutron star without being disrupted.

Although it is not clear whether a hypothetical neutron star would be able to pull this off, it is interesting to note how close their model comes to the stationary-beam ETI beacon model described at the end of chapter 7. The only differences are the following: a) in the ETI model these modulating standing wave fields are engineered rather than produced by nature, b) these fields do not rotate with the star, but are artificially projected near the star's surface, and c) the star need not necessarily be a neutron star, but could also be a white-dwarf-sized X-ray star.

Interestingly, this unusual pulsar is the Vela pulsar's nearest pulsar neighbor. The two are separated by only 11 degrees of arc and lie about the same distance from the Sun, PSR 0826–34 lying only 4 percent farther away. Their placement relative to one another shows no particularly significant alignment as was observed between the Crab pulsar and PSR 0525+21. But this coincidence is worth noting.

Mode Switching

As described in chapter 7, mode switching is one of the most perplexing properties of pulsars. Pulsars that exhibit this behavior have more than one stable pulsation mode, each with its own time-averaged pulse profile and unique pulse modulation characteristics. Such pulsars will suddenly switch from one mode to another without showing any change in either their primary period or their period derivative. As of the time of writing, 7 pulsars have been identified as mode switchers, two others being possible additional candidates.

One interesting mode switcher, PSR 1237+25, exhibits two stable modes, each having five components to its time-averaged pulse profile (figure A.5). The two modes are designated as *normal* and *abnormal*. The pulsar usually transmits in the normal mode, but about every few hours it flips into its abnormal mode and stays there for several minutes before flipping back. When pulsing in its normal mode, the pulsar exhibits very unusual pulse-ordering behavior. Pulses alternately emerge in component-1 and component-5 of its five-component time-averaged profile, pulses reappearing in each component with a characteristic period P_3 of 2.7 seconds, equal to about twice the pulsar's primary period of 1.382449 seconds. But what is even more unusual is that pulses emerging in component-1 are observed to subsequently split into two separate pulse trains that drift in opposite directions, one toward the leading edge of the component and the other toward the trailing edge. When the pulsar switches to its abnormal mode, this unusually complex correlated multi-component modulation almost totally disappears. Upon returning to the normal mode, the periodic behavior is faithfully remembered and resumed. Such intricate signal ordering gives very good reason to consider that these transmissions may have an ETI origin.

Another mode switcher, PSR 0329+54, exhibits as many as three abnormal modes in addition to its normal mode! Moreover, this pulsar has differing numbers of abnormal modes available to it depending on the particular radio frequency that the pulsar is viewed at. As seen in figure A.6, abnormal mode A is present at frequencies of 0.41 gigahertz and from 5 to 14.8 gigahertz; abnormal mode B appears at a frequency of 0.83 gigahertz; both abnormal modes A and B are detected at 2.7 gigahertz; and all three abnormal modes are present at 1.4 gigahertz. As mentioned in chapter 7, we are led to question whether it is just a coincidence that all three switching modes are present at the 1.4 gigahertz neutral-hydrogen emission frequency that SETI astronomers deem to be the most likely frequency to be used in narrow-band ETI radio communication.

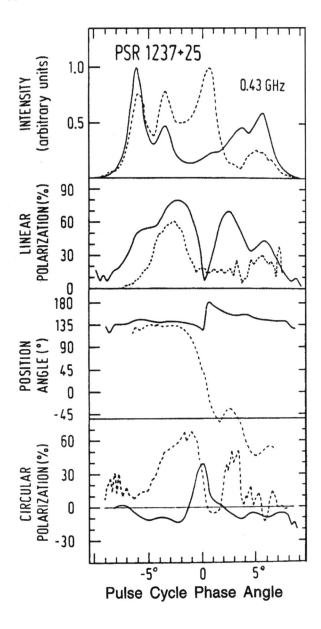

Figure A.5. Intensity and polarization properties of the time-averaged pulse profile for PSR 1237+25 observed at a radio frequency of 430 megahertz. Normal mode: solid line; abnormal mode: dashed line (Bartel et al., Astrophysical Journal, *figure 7).*

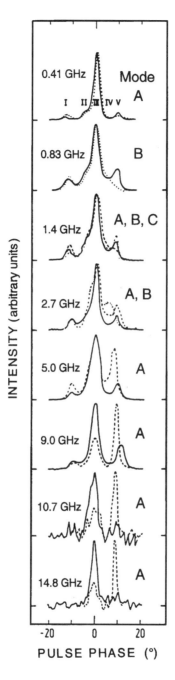

Figure A.6. Time-averaged pulse profiles for PSR 0329+54 at each of eight frequencies. The normal mode is shown as a solid line. The three abnormal modes, designated A, B, and C, are shown as dashed and dotted lines (Bartel et al., Astrophysical Journal, *figure 3).*

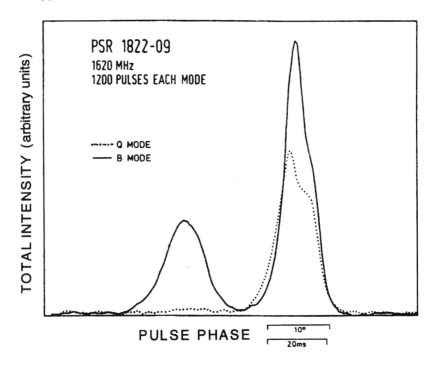

Figure A.7. Quiescent and burst mode mean profiles for the main pulse component of PSR 1822–09 at 1.62 gigahertz (Fowler et al., Astronomy and Astrophysics, figure 3).

PSR 1822–09 is another unique mode-switching pulsar. About every five minutes, it switches between a quiescent mode (Q mode) and a burst mode (B mode); see figure A.7. The quiescent mode is characterized by a single-component main pulse followed about half a pulsation cycle later by a weak interpulse (not shown in the diagram). When the pulsar switches to its burst mode, the main peak almost doubles its intensity and is accompanied by a new "precursor peak" of lesser intensity. Astronomers have discovered that during the quiescent mode, pulses in the interpulse and main pulse are correlated, implying that the emission processes in the two peaks are causally linked. However, contrary to expectation, when the pulsar switches to its burst mode, the interpulse does not increase in intensity like the main pulse, but instead vanishes below the threshold of detection! This paradox poses quite a problem to theoreticians who seek a logical natural cause for this unusual behavior. Such paradoxical behavior, though, would certainly be plausible if this mode of behavior were one of many techniques used by extraterrestrials to foil our attempts to pass off their message as a natural phenomenon.

The quiescent and burst modes of this pulsar also exhibit very different pulse-drifting behavior. When the pulsar is in its quiescent mode, pulses drift from the leading-to-trailing edge of its main peak with a relatively well-defined period P_3 equal to about 40 times the pulsar's primary period. The drift rate varies such that the pulses accelerate as they approach the center of the main pulse profile and then decelerate as they approach the profile's trailing edge. When the pulsar switches to the burst mode, pulse drifting in the main pulse ceases, but appears sporadically 5 to 10 percent of the time in the precursor pulse with a period equal to 11 times the primary period.

Mode Switching with Pulse Quantization. Another interesting pulsar, PSR 0031–07, exhibits mode switching among three modes: A, B, and C, each exhibiting pulse drifting behavior and each with its own drift rate. The drift rates for modes A, B, and C are respectively 1.7±0.2, 3.2±0.6, and 5.3±1.1 degrees of phase per primary pulse period, suggesting a quantized ratio relationship of 1:2:3. Thus, not only are these pulses ordered so that they drift in phase in a regular manner, but also this drifting process, in turn, is ordered to allow only drifting rates that bear a certain integer relation to one another. Moreover, despite these three different pulse drifting speeds (three different profile scan periods, P_3), in all three modes the interval between successive pulses, period P_2, stays fixed at a value equal to 6 percent of the pulsar's primary period. While these ordered relationships make little sense from the standpoint of contemplating a possible natural cause, it would be quite plausible if this pulsar, like many others, is an extraterrestrial communication beacon.

Mode Switching with Pulse Nulling. Besides these complexities, PSR 0031–07 exhibits unusual interruptions in its pulsed emissions (see figure A.8). Rather than transmitting a steady train of pulses, it emits bursts of pulses consisting of several dozen to as many as 100 pulses. These bursts are separated by blank intervals of varied lengths made up of null pulses that are at least a hundredfold weaker than the burst mode pulses and often are totally undetectable. Null intervals may span anywhere from a few to as many as 100 or more pulse periods. Most bursts from this pulsar begin abruptly with a pair of high-intensity pulses, one positioned close to the leading edge of the time-averaged pulse profile and the other positioned close to its trailing edge. The pulses that follow then go on to drift toward the envelope's leading edge at the characteristic drift rate of the mode the pulsar happens to be switched to at that time.

Mode Switching with Pulse Grammar. About 80 percent of the bursts from PSR 0031–07 are dominated by Mode B pulse drifting. The remaining 20 percent have been the subject of considerable interest. They consist of two drifting modes that occur consecutively, either A followed by B or B followed by C. In such bursts, the pulsar might make several pulse scans across the time-averaged pulse profile at one mode (e.g., A) and then, right in the middle of the burst, suddenly switch to an alternate mode with a different pulse drift rate (e.g., B); see figure A.9. About three fourths of these dual-mode bursts are composed of A and B mode sequences and about one fourth are composed of B and C mode sequences. A and C modes *never occur together in the same burst; nor do they occur alone like B mode bursts.* Thus, not only are pulse drift rates quantized in this pulsar, but also the process of switching from one quantized drift rate to another appears to be highly structured, being governed by a kind of *pulse mode grammar.*

Mode Switching with Pulse Phase Memory. PSR 0031–07 exhibits yet another ordering effect. During a nulling period, it somehow "remembers" the phase positions its pulses had at the time its emission

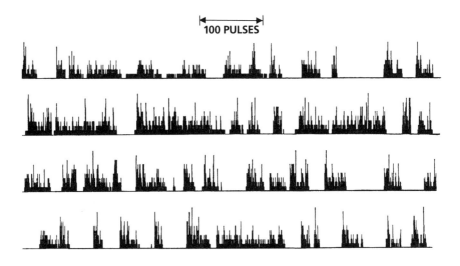

Figure A.8. A continuous record of 3,800 pulses from pulsar PSR 0031–07 (Huguenin et al., Astrophysical Journal, *figure 1).*

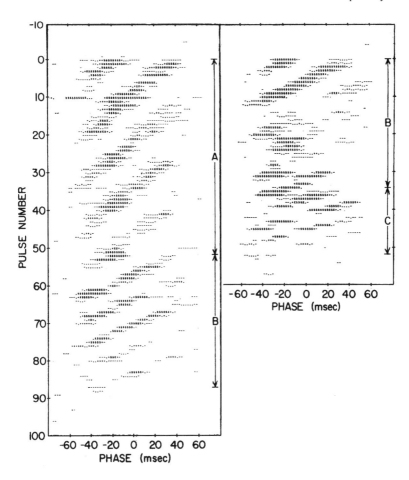

Figure A.9. A succession of pulses received from PSR 0031–07 showing linearly drifting pulse trains. The three drift modes are marked in the right margins as A, B, and C (adapted from Huguenin et al., Astrophysical Journal, *figure 4).*

shut off so that when the pulsar turns on again with a new burst, its pulses have not moved appreciably from their former phase positions. It is as if pulse drifting "freezes" while nulling is in progress, and then recommences when nulling ends. Careful studies of two single mode pulsars that exhibit pulse nulling (PSR 0809+74 and PSR 0818–13) indicate that pulses are present during nulls, although at very low intensity. In this case, the pulse drifting does not freeze, but rather continues to drift, although at a very reduced rate.[5] Consequently, this type of nulling could be classified as a kind of mode-switching phenomenon.

Mode Switching with Mode Memory. The mode-switching pulsar PSR 2319+60 also emits its pulses in bursts separated by null periods that occur about 30 percent of the time.[6] It has three modes designated A, B, and Abn (an abnormal mode), each of which is characterized by a different pulse drift rate. Like PSR 0031–07, its process of switching from one mode to another is governed by rules. During a burst, mode switching is allowed to occur either as A → B, B → Abn, or A → B → Abn. No reverse transitions have been observed, such as B → A or A → Abn. Moreover, the mode transition rules often appear to span null periods. For example, the forward transition, A → null → B, occurs much more frequently than the reverse, B → null → A. It is as if this mode-switching pulsar remembers what mode it was in before the null period began (e.g., mode A) so that it can adopt an allowed mode after the null period ends (e.g., mode B).

The various types of complex signal ordering reviewed above would be the kind of thing we might expect if pulsars have been fabricated by intelligent beings. Such varied kinds of signal ordering could be their attempt to convince us that pulsars are indeed artifactual beacons.

APPENDIX B

PARTICLE BEAM–
COMMUNICATOR
LUMINOSITY

The radio luminosity coming from the "low-tech" particle-beam communicator described in chapter 7 may be calculated using the following equation:

$$L = 4\pi d^2 \ (W/P_1) \ (S/8\gamma^3) \ \Delta\nu, \tag{1}$$

where d is the distance of the source from the Earth, W is the pulse width at half maximum intensity, P_1 is the primary pulse period, S is the radiation flux density of the source at a given frequency as seen by an Earth observer, γ is the Lorentz factor of the radiating electrons due to their relativistic velocity v (i.e., $\gamma = 1/\sqrt{1-v^2/c^2}$), and $\Delta\nu$ is the bandwidth under consideration. The quantity $8\gamma^3$ in the above equation is the factor by which we must divide S, the observed radiation flux density, to obtain the actual radiation flux density of the source to adjust for relativistic beaming.

This example assumes that the electrons in the communicator's particle beam have a kinetic energy of 50 billion electron volts, which is equivalent to a Lorentz factor of $\gamma = 2 \times 10^5$. Also suppose that $S = 8 \times 10^{-24}$ ergs/second/centimeter^2/Hertz at a frequency of 400 megahertz; which is equivalent to the flux density observed from the Crab pulsar. Also let d = 6600 light-years = 6.6×10^{21} centimeters, the distance to the Crab pulsar, and let W/P = 0.42, the width-to-pulse-period ratio for the Crab pulsar. For a bandwidth of $\Delta\nu = 400$ megahertz, relation

(1) predicts a luminosity of 10^{13} ergs/sec = 1 megawatt. By comparison the 20-trillion-electron-volt superconducting supercollider the United States had planned to build was expected to achieve a particle power flux 300 times this amount.

This estimates just the energy stored in the electron beam. The actual power drawn by the accelerator to achieve this electron energy flux will be substantially greater. With a one percent acceleration efficiency, the total power requirement amounts to 100 megawatts, the amount supplied by a small nuclear power plant.

The Crab pulsar is one of the few that emit pulses at optical, X-ray, and gamma ray energies. In fact, in this high-energy portion of the electromagnetic spectrum it radiates 10^4 to 10^5 times the amount of energy that it emits in the radio region of its spectrum. These higher luminosities could be attained by accelerating the beam's electrons to 2 trillion electron volts ($\gamma = 8 \times 10^6$).

NOTES

1. The Pulsar Enigma

1. W. Sullivan, *Black Holes* (Garden City, NY: Anchor Press, 1979), 123.
2. A. Hewish, S. J. Bell, J. D. H. Pilkington, P. F. Scott, and R. A. Collins, "Observation of a rapidly pulsating radio source," *Nature* 217 (1968): 709–13.
3. *New York Post,* as cited in R. Collyns, *Did Spacemen Colonize the Earth?* (London: Pelham Books, 1974), 231.
4. R. Jastrow and M. H. Thompson, *Astronomy: Fundamentals and Frontiers* (New York: John Wiley & Sons, 1977), 198.
5. T. Gold, "Rotating neutron stars as the origin of the pulsating radio sources," *Nature* 218 (1968): 731–32.
6. C. Misner, K. Thorne, and J. A. Wheeler, *Gravitation* (San Francisco: Freeman and Co., 1973), 628.
7. *Winnipeg Free Press,* November 27, 1974.

2. A Galactic Message

1. P. A. LaViolette, "Evidence that radio pulsars may be artificial beacons of ETI origin," 195th American Astronomical Society meeting, Atlanta, Georgia, January 2000.
2. Pulsar data is taken from ftp://pulsar.princeton.edu/pub/catalog/.
3. D. H. Clark and J. L. Caswell, "A study of Galactic supernova remnants based on Molonglo-Parkes observational data," *Monthly Notices of the Royal Astronomical Society* 174 (1996): 267–305, fig. 8.
4. D. Backer et al., "A millisecond pulsar," *Nature* 300 (1982): 615–18.
5. "Newly discovered pulsar is 20 times faster than Crab pulsar," *Physics Today,* March 1983, pp. 19–21.

6. G. Hobbs et al., "A statistical study of 233 pulsar proper motions," *Monthly Notices of the Royal Astronomical Society* 360 (2005): 974–92.

7. P. A. LaViolette, *Earth Under Fire: Humanity's Survival of the Ice Age* (Rochester, Vt.: Bear & Co., 2005).

8. A. S. Fruchter, D. R. Stinebring, and J. H. Taylor, "A millisecond pulsar in an eclipsing binary," *Nature* 333 (1988): 237–39.

9. R. Preston, "The eclipsing death star," *Discovery,* August 1988: 41–46.

10. Hobbs et al., "A statistical study of 233 pulsar proper motions."

11. A. L. Webre, *Exopolitics: Politics, Government, and Law in the Universe* (Vancouver, B.C.: Universe Books, 2005).

12. J. W. Hessels et al., "A 20 cm search for pulsars in globular clusters with Arecibo and the GBT," in *Young Neutron Stars and Their Environments,* edited by F. Camilo and B. Gaensler, IAU Symp., vol. 218, 2004; arxiv.org/abs/astro-ph/0402182.

13. J. W. Hessels, personal communication, 2005.

3. The Galactic Network

1. T. T. Brown, "How I control gravity," *Science and Invention,* August 1929.

2. P. A. LaViolette, *Subquantum Kinetics: A Systems Approach to Physics and Cosmology* (Niskayuna, N.Y.: Starlane Publications, 2003), ch. 11.

3. P. Cornille, "Review of the application of Newton's third law in physics," *Progress in Energy and Combustion Science* 25 (1999): 161–210.

4. Gravity Research Group, "Electrogravitics Systems: An examination of electrostatic motion, dynamic counterbary, and barycentric control," Report GRG 013/56, London: Aviation Studies (International) Ltd., Special Weapons Study Unit, February 1956. (Declassified report: Wright Patterson Air Force Base ID number 3-1401-00034-5879.)

5. P. A. LaViolette, "The U.S. Antigravity Squadron," in *Electrogravitics Systems: Reports on a New Propulsion Methodology,* ed. T. Valone (Washington, D.C.: Integrity Research Institute, 1994).

6. P. A. LaViolette, "An introduction to subquantum kinetics," *International Journal of General Systems* 11 (1985): 281–345.

7. LaViolette, *Subquantum Kinetics.*

8. E. Podkletnov and G. Modanese, "Impulse gravity generator based on charged $YBa_2Cu_3O_{7-y}$ superconductor with composite crystal structure," August 2001. Eprint: arXiv.org/abs/physics/0108005.

9. E. Podkletnov and G. Modanese, "Investigation of high voltage discharges in low pressure gases through large ceramic superconducting electrodes," *Journal of Low Temperature Physics* 132 (2003): 239–59. Eprint: arXiv.org/abs/physics/0209051.

10. E. Podkletnov and G. Modanese, "Antigravity propulsion comes out of the closet," *Janes Defense Weekly,* July 31, 2002.

11. N. Cook, "Airpower electric," *Janes Defense Weekly,* July 24, 2002.

12. E. Podkletnov, personal communication, 2003.

13. Ibid.

14. T. K. Ishii and G. C. Giakos, "Radio Messages Faster Than Light," *Microwaves & RF* 30 (August 1991): 114–19.

15. T. K. Ishii and G. C. Giakos, "Rapid Pulsed Microwave Propagation," *IEEE Microwave & Guided Wave Letters* 1, no. 12 (December 1991): 374–75.

16. A. Enders and G. Nimtz, "On superluminal barrier traversal," *Physical Review E* 48 (1993): 632.

17. N. Hawkes, "Going faster than light," *London Times,* April 3, 1995, p. 14.

18. W. M. Cady, "An investigation relative to Thomas Townsend Brown," Office of Naval Research, Pasadena, Calif., June 1952, p. 1.

19. T. T. Brown, "Electrogravitational communication system," patent disclosure, September 1953; available at http://www.soteria.com.

20. Intel, "Towards flight without stress or strain . . . or weight," *Interavia* 11, no. 5 (1956): 373–74.

21. J. Dea, "Instantaneous interactions," *Proceedings of the 1986 International Tesla Symposium* (Colorado Springs, Colo.: International Tesla Society, 1986), 4–39.

4. The Galactic Imperative

1. P. A. LaViolette, *Earth Under Fire: Humanity's Survival of the Ice Age* (Rochester, Vt.: Bear & Co., 2005).

2. P. A. LaViolette, *Genesis of the Cosmos: The Ancient Science of Continuous Creation* (Rochester, Vt.: Bear & Co., 2004).

3. Ibid.

4. P. A. LaViolette, "Galactic explosions, cosmic dust invasions, and climatic change," Ph.D. dissertation, Portland State University, 1983;

CD-ROM update, *Galactic Superwaves and Their Impact on the Earth,* 2005.

5. P. A. LaViolette, "Cosmic-ray volleys from the Galactic center and their recent impact on the Earth environment," *Earth, Moon, and Planets* 37 (1987): 241–86.

6. G. C. Bower, H. Falcke, and D. C. Backer, "Circular polarization in Sagittarius A*," 195th American Astronomical Society meeting, Atlanta, January 2000.

7. LaViolette, "Galactic explosions . . . ," ch. 3.

8. Ibid., ch. 4, 5.

9. LaViolette, *Earth Under Fire,* ch. 3, 10.

10. C. U. Hammer, H. B. Clausen, and C. C. Langway Jr., "50,000 years of recorded global volcanism," *Climatic Change* 35 (1997): 1–15.

11. P. A. LaViolette, "Solar cycle variations in ice acidity at the end of the last ice age: Possible marker of a climatically significant interstellar dust incursion," *Planetary & Space Science* 53, no. 4 (2005): 385–93. Eprint: arXiv.org/abs/physics/0502019.

12. P. A. LaViolette, "Evidence for a global warming at the Termination I boundary and its possible extraterrestrial cause," 2005. Eprint: arXiv.org/abs/physics/0503158.

13. H. A. Zook, J. B. Hartung, and D. Storzer, "Solar flare activity: Evidence for large-scale changes in the past," *Icarus* 32 (1977): 106–26.

14. T. Gold, "Apollo II observations of a remarkable glazing phenomenon on the lunar surface," *Science* 165 (1969): 1345–49.

15. B. S. Boeckl, "A depth profile of ^{14}C in the lunar rock 12002," *Earth and Planetary Science Letters* 16 (1972): 269–72.

16. LaViolette, "Evidence for a global warming."

17. K. A. Hughen et al., "Deglacial changes in ocean circulation from an extended radiocarbon calibration," *Nature* 391 (1998): 65–68.

18. LaViolette, *Earth Under Fire,* ch. 7.

19. P. A. LaViolette, *Genesis of the Cosmos: The Ancient Science of Continuous Creation* (Rochester, Vt.: Park Street Press, 2004).

20. LaViolette, *Earth Under Fire,* ch. 2.

21. LaViolette, "Evidence for a global warming."

22. LaViolette, "Galactic explosions . . . ," ch. 4, 5.

23. LaViolette, *Earth Under Fire,* ch. 3, 10.

24. Ibid., ch. 1.

25. LaViolette, *Genesis of the Cosmos,* ch. 9.

26. R. L. Thompson, *Alien Identities: Ancient Insights into Modern UFO Phenomena* (San Diego, Calif.: Govardhan Hill Publishing, 1993).

27. Plutarch, "Lucullus," in *Plutarch's Lives,* vol. 2, trans. B. Perrin (Cambridge, Mass.: Harvard University Press, 1968), 495–97.

5. Superwave Warning Beacons

1. W. H. Tucker, "Supernova in the sail," *Star and Sky* 2, no. 1 (1980): 36.

2. G. Michanowsky, *The Once and Future Star* (New York: Barnes and Noble, 1979).

3. P. A. LaViolette, *Earth Under Fire: Humanity's Survival of the Ice Age* (Rochester, Vt.: Bear & Co., 2005), ch. 7.

4. J. H. Taylor and D. R. Stinebring, "Recent progress in the understanding of pulsars," *Annual Reviews of Astronomy and Astrophysics* 24 (1986): 303.

5. Y. Gupta, B. J. Rickett, and A. G. Lyne, "Refractive Interstellar Scintillation in Pulsar Dynamic Spectra," *Monthly Notices of the Royal Astronomical Society* 269 (1994): 1035.

6. A. De Luca, R. P. Mignani, and P. A. Caraveo, "The Vela pulsar (PSR B0833-45) proper motion revisited with HST astrometry," *Astronomy and Astrophysics* 354 (2000): 1011–13.

7. G. F. Bignami and P. A. Caraveo, "On the birthplace of PSR 0833-45: Or, is the Vela pulsar associated with the Vela SNR?" *Astrophysical Journal* 325 (1988): L5–L7.

8. V. Trimble, "Motions and Structure of the Filamentary Envelope of the Crab Nebula," *Astronomical Journal* 73 (1968): 535–47.

9. J. H. Taylor, R. N. Manchester, and A. G. Lyne, "Catalog of 558 pulsars," *Astrophysical Journal Supplement Series* 88 (1993): 553.

10. R. Isaacman, "NP0532 and a hole in the Crab Nebula," *Nature* 268 (1977): 317–18.

11. P. A. LaViolette, "Galactic explosions, cosmic dust invasions and climatic change," Ph.D. dissertation, Portland State University, 1983, ch. 5.

12. LaViolette, *Earth Under Fire,* pp. 287–88.

13. LaViolette, "Galactic explosions . . . ," ch. 5.

14. LaViolette, *Subquantum Kinetics: A Systems Approach to Physics and Cosmology* (Niskayuna, N.Y.: Starlane Publications, 2003), 225–26.

15. LaViolette, "The planetary-stellar mass-luminosity relation: Possible evidence of energy nonconservation?" *Physics Essays* 5, no. 4 (1992): 536–44.

16. W. C. Erickson et al., "Very long baseline interferometer observations of Taurus A and other sources at 121.6 MHz," *Astrophysical Journal* 177 (1972): 101.

17. LaViolette, "Galactic explosions . . . ," ch. 5.

18. LaViolette, *Earth Under Fire,* ch. 10.

19. Ibid.

20. LaViolette, "Galactic explosions . . . ," ch. 5.

21. LaViolette, *Earth Under Fire,* ch. 10.

22. LaViolette, "Galactic explosions . . . ," ch. 5.

23. G. D. Schmidt, J. Angel, and E. Beaver, "The small-scale polarization of the Crab Nebula," *Astrophysical Journal* 227 (1979): 106–13.

24. W. J. Cocke, M. Disney, G. Muncaster, and T. Gehrels, "Optical polarization of the Crab Nebula pulsar," *Nature* 227 (1970): 1327–29.

25. LaViolette, "Galactic explosions . . . ," ch. 5.

26. P. A. LaViolette, "Cosmic-ray volleys from the Galactic center and their recent impact on the Earth environment," *Earth, Moon, and Planets* 37 (1987): 241–86.

27. LaViolette, *Earth Under Fire,* 283.

28. S. P. Reynolds and D. C. Ellison, "Electron acceleration in Tycho's and Kepler's supernova remnants: Spectral evidence of Fermi shock acceleration," *Astrophysical Journal* 399 (1992): L75–L78.

29. C. Heiles, D. B. Campbell, and J. M. Rankin, "Pulsar NP 0532: Properties and systematic polarization of individual strong pulses at 430 MHz," *Nature* 226 (1970): 529–31.

30. J. M. Cordes et al., "The brightest pulses in the universe: Multifrequency observations of the Crab pulsar's giant pulses," *Astrophysical Journal* 612 (2004): 375–88.

31. S. Johnston et al., "High time-resolution observations of the Vela pulsar," *Astrophysical Journal* 549 (2001): L149.

32. M. Kramer, S. Johnston, and W. van Straten, "High-resolution single-pulse studies of the Vela pulsar," *Monthly Notices of the Royal Astronomical Society* 334 (2002): 523.

33. A. G. Lyne, R. S. Pritchard, F. Graham-Smith, and F. Camilo, "Very low braking index for the Vela pulsar," *Nature* 381 (1996): 497–98.

34. A. G. Lyne and F. Graham-Smith, *Pulsar Astronomy* (London: Cambridge University Press, 1998), 63.

35. C. L. Bhat, M. L. Sapru, and C. L. Kaul, "A nonrandom component in cosmic rays of energy greater than or equal to 10 to the 14th eV," *Nature* 288 (1980): 146–49.

36. S. D. Hyman et al., "A new radio detection of the bursting source GCRT J145–3009," 2005, eprint: www.arxiv.org/abs/astro-ph/0508264.

6. Sky Maps of a Celestial Disaster

1. P. A. LaViolette, "Galactic explosions, cosmic dust invasions, and climatic change," Ph.D. dissertation, Portland State University, 1983, pp. 291, 304.

2. P. A. LaViolette, *Earth Under Fire: Humanity's Survival of the Ice Age* (Rochester, Vt.: Bear & Co., 2005).

3. Ibid., 36–39.

4. Ibid., 61–66.

5. Ibid., ch. 2.

6. Ibid., 76–88.

7. Ibid., 160, 171–72.

8. Ibid., 235–38.

9. T. R. Gull and R. A. Fesen, "Deep optical imagery of the Crab Nebula's jet," *Astrophysical Journal* 260 (1982): L75–L78.

10. Ibid.

11. R. A. Fesen and B. Staker, "The structure and motion of the Crab nebula jet," *Monthly Notices of the Royal Astronomical Society* 263 (1993): 69–74.

12. T. Velusamy, "Radio detection of a jet in the Crab Nebula," *Nature* 308 (1984): 251–52.

13. J. McMoneagle, *The Ultimate Time Machine* (Charlottesville, Va.: Hampton Roads, 1999).

14. J. A. West, *Serpent in the Sky: The High Wisdom of Ancient Egypt* (Wheaton, Ill.: Quest, 1993).

15. A. S. Fruchter et al., "The eclipsing millisecond pulsar PSR 1957+20," *Astrophysical Journal* 351 (1990): 642–50.

7. Natural or Artificial

1. A. Wolszczan and D. A. Frail, "A planetary system around the millisecond pulsar PSR 1257+12," *Nature* 355 (1992): 145–47.

2. M. Demianski, and M. Prószynski, "Does PSR 0329+54 have companions?" *Nature* 282 (1979): 383–85.

3. A. S. Fruchter et al., "The eclipsing millisecond pulsar PSR 1957+20," *Astrophysical Journal* 351 (1990): 642–50.

4. D. A. Moffet and T. H. Hankins, "Multifrequency radio observations of the Crab pulsar," *Astrophysical Journal* 468 (1996): 779–83.

5. T. H. Hankins and J. M. Cordes, "Interpulse emission from pulsar 0950+08: How many poles?" *Astrophysical Journal* 249 (1981): 241.

6. S. Johnston et al., "Discovery of a very bright, nearby binary millisecond pulsar," *Nature* 361 (1995): 613–15.

7. D. Backer et al., "A millisecond pulsar," *Nature* 300 (1982): 615–20.

8. J. Winn, "The life of a neutron star," *Sky & Telescope,* July 1999: 34.

9. N. Bartel, D. Morris, W. Sieber, and T. H. Hankins, "The mode-switching phenomenon in pulsars," *Astrophysical Journal* 258 (1982): 777.

10. Ibid.

11. G. R. Huguenin, J. H. Taylor, and T. H. Troland, "The radio emission from pulsar MP 0031-07," *Astrophysical Journal* 162 (1970): 727–35.

12. A. V. Filippenko and V. Radhakrishnan, "Pulsar nulling and drifting subpulse phase memory," *Astrophysical Journal* 263 (1982): 828–34.

13. J. H. Taylor, R. N. Manchester, and G. R. Huguenin, "Observations of pulsar radio emission. I. Total-intensity measurements of individual pulses," *Astrophysical Journal* 195 (1975): 513–28.

14. J. M. Dawson, "Plasma particle accelerators," *Scientific American,* March 1989: 54–61.

15. P. A. LaViolette, *Subquantum Kinetics: A Systems Approach to Physics and Cosmology* (Niskayuna, N.Y.: Starlane Publications, 2003), ch. 10.

16. B. Eastlund, "Method and Apparatus for Altering a Region in the Earth's Atmosphere, Ionosphere, and/or Magnetosphere," U.S. patent No. 4,686,605, August 11, 1987.

8. Force Field–Beaming Technology

1. T. E. Bearden, *Fer-de-Lance: A Briefing on Soviet Scalar Electromagnetic Weapons* (Ventura, Calif.: Tesla Book Co., 1986).

2. H. Mason, "Bright Skies: Top-Secret Weapons Testing?" *Nexus,* April–May 1997: 45.

3. This firsthand account was related directly to the author by these individuals, who wanted to be known by their first names only.

4. Mason, "Bright Skies . . . ," 41–47, 78.

5. V. V. Shkunov and B. Y. Zel'dovich, "Optical phase conjugation," *Scientific American* 253 (December 1985): 54–59.

6. D. M. Pepper, "Applications of optical phase conjugation," *Scientific American* 254 (January 1986): 74–83.

7. Bearden, *Fer-de-Lance*.

8. T. E. Bearden, "Soviet phase conjugate weapons," Bulletin no. 308, Committee to Restore the Constitution, January 1988.

9. P. A. LaViolette, *Subquantum Kinetics: A Systems Approach to Physics and Cosmology* (Niskayuna, N.Y.: Starlane Publications, 2003), ch. 6.

10. G. Vassilatos, *Secrets of Cold War Technology: Project HAARP and Beyond* (Bayside, Calif.: Borderland Sciences, 1996), 26–33.

11. Ibid., 51, 55.

12. Ibid., 87.

13. P. Delgado and C. Andrews, *Circular Evidence* (London: Bloomsbury Publishing, 1989), 127–28.

14. Ilyes, *An Hypothesis: The Transmission of a Crop Circle,* 1996. On Web at www.cropcircleconnector.com/ilyes/Abouthy.html.

15. F. Silva, "Music in the Fields," *Atlantis Rising,* no. 14 (1998): 42–43.

16. W. C. Levengood, "Anatomical Anomalies in Crop Formation Plants," *Physiologia Plantarum* 92 (1994): 356–63.

17. R. Russell, "Report on preliminary results of electrostatic energy testing in crop formations," Midwest Research, Aurora, Colorado, August 1999.

18. P. Delgado and C. Andrews, *Circular Evidence,* 158.

19. Ilyes, *An Hypothesis: The Transmission of a Crop Circle.*

20. D. Higbee, "Crop circles: Real or hoax?" Web posting at users1.ee.net/pmason/crop-circles.html.

21. P. Delgado and C. Andrews, *Circular Evidence,* 115.

22. Ilyes, *An Hypothesis: Transmission of a Crop Circle,* excerpt from story by R. Dutton, printed in spring 1996 issue of *Circular.*

23. Higbee, "Crop circles: Real or hoax?"

24. P. A. Sturrock et al., "Physical evidence related to UFO reports," *Journal of Scientific Exploration* 12 (1998): 179–229.

25. P. Hill, *Unconventional Flying Objects: A Scientific Analysis* (Charlottesville, Va.: Hampton Roads Publishing, 1995), 98–116.

26. Ibid., 105.

27. C. Lorenzen and J. Lorenzen, *UFO: The Whole Story* (New York: Signet, 1968), 97.

28. B. Steiger, *Alien Meetings* (New York: Ace, 1978).
29. G. Zeitlin, "Are pulsar signals evidence of astro-engineered signalling systems?" *New Frontiers in Science* 1 (Summer 2002).
30. A. Lyne and F. Graham-Smith, *Pulsar Astronomy* (London: Cambridge University Press, 1998), 80.

Appendix A: Ordered Complexity

1. J. H. Taylor, R. N. Manchester, and G. R. Huguenin, "Observations of pulsar radio emission. I. Total-intensity measurements of individual pulses," *Astrophysical Journal* 195 (1975): 513–28.
2. J. D. Biggs, P. M. McCulloch, P. A. Hamilton, R. N. Manchester, and A. G. Lyne, "A study of PSR 0826-34—a remarkable pulsar," *Monthly Notices of Royal Astronomical Society* 215 (1985): 281–94.
3. Ibid., 292.
4. Ibid., 281.
5. J. H. Taylor and D. Stinebring, "Recent progress in the understanding of pulsars," *Annual Reviews of Astronomy and Astrophysics* 24 (1986): 285–327.
6. G. A. E. Wright and L. A. Fowler, "Mode-changing and quantized subpulse drift-rates in pulsar PSR 2319+60," *Astronomy and Astrophysics* 101 (1981): 356–61.

BIBLIOGRAPHY

Backer, D. C. "Pulsar fluctuation spectra and the generalized drifting subpulse phenomenon." *Astrophysical Journal* 182 (1973): 245–76.

Backer, D. C., et al. "A millisecond pulsar." *Nature* 300 (1982): 615–20.

Bartel, N., D. Morris, W. Sieber, and T. H. Hankins. "The mode-switching phenomenon in pulsars." *Astrophysical Journal* 258 (1982): 776–89.

Bearden, T. E. *Fer-de-Lance: A Briefing on Soviet Scalar Electromagnetic Weapons.* Ventura, Calif.: Tesla Book Co., 1986.

———. "Soviet phase conjugate weapons." Bulletin no. 308, Committee to Restore the Constitution, January 1988.

Beer, J., et al. "^{10}Be measurements on polar ice: Comparison of Arctic and Antarctic records." *Nuclear Instruments and Methods in Physics Research,* B29 (1987): 203–206.

———. "^{10}Be peaks as time markers in polar ice cores." In *The Last Deglaciation: Absolute and Radiocarbon Chronologies,* NATO ASI Series, vol. 12, 140–53. Heidelberg: Springer-Verlag, 1992.

Bhat, C. L., M. L. Sapru, and C. L. Kaul. "A nonrandom component in cosmic rays of energy greater than or equal to 10 to the 14th eV." *Nature* 288 (1980): 146–49.

Biggs, J. D., P. M. McCulloch, P. A. Hamilton, R. N. Manchester, and A. G. Lyne. "A study of PSR 0826-34—a remarkable pulsar." *Monthly Notices of Royal Astronomical Society* 215 (1985): 281–94.

Bignami, G. F., and P. A. Caraveo. "On the birthplace of PSR 0833-45: Or, is the Vela pulsar associated with the Vela SNR?" *Astrophysical Journal* 325 (1988): L5–L7.

Boeckl, B. S. "A depth profile of 14C in the lunar rock 12002." *Earth and Planetary Science Letters* 16 (1972): 269–72.

Bower, G., H. Falcke, and D. C. Backer. "Circular polarization in Sagittarius A*." 195th American Astronomical Society meeting, Atlanta, January 2000.

Briggs, J., and D. Peat. *Turbulent Mirror.* San Francisco: Harper & Row, 1989.

Brown, T. T. "How I control gravity." *Science and Invention,* August 1929.

———. "Electrogravitational communication system." Patent disclosure, September 1953; available at http://www.soteria.com.

Cady, W. M. "An investigation relative to Thomas Townsend Brown." Office of Naval Research, Pasadena, Calif., June 1952.

Clark, D. H., and J. L. Caswell. "A study of Galactic supernova remnants based on Molonglo-Parkes observational data." *Monthly Notices of the Royal Astronomical Society* 174 (1996): 267–305.

Cocke, W. J., M. Disney, G. Muncaster, and T. Gehrels. "Optical polarization of the Crab Nebula pulsar." *Nature* 227 (1970): 1327–29.

Collyns, R. *Did Spacemen Colonize the Earth?* London: Pelham Books, 1974.

Cook, N. "Antigravity propulsion comes out of the closet." *Janes Defense Weekly,* July 31, 2002.

———. "Airpower electric." *Janes Defense Weekly,* July 24, 2002.

Cordes, J. M., et al. "The brightest pulses in the universe: Multifrequency observations of the Crab pulsar's giant pulses." *Astrophysical Journal* 612 (2004): 375–88.

Cornille, P. "Review of the application of Newton's third law in physics." *Progress in Energy and Combustion Science* 25 (1999): 161–210.

Dawson, J. M. "Plasma particle accelerators." *Scientific American,* March 1989: 54–61.

De Luca, A., R. P. Mignani, and P. A. Caraveo, "The Vela pulsar (PSR B0833-45) proper motion revisited with HST astrometry." *Astronomy and Astrophysics* 354 (2000): 1011–13.

Dea, J. "Instantaneous interactions." In *Proceedings of the 1986 International Tesla Symposium,* 4–39, Colorado Springs: International Tesla Society, 1986.

Delgado, P., and C. Andrews. *Circular Evidence.* London: Bloomsbury Publishing, 1989.

Demianski, M., and M. Prószynski. "Does PSR 0329+54 have companions?" *Nature* 282 (1979): 383–85.

Dewey, R. J., J. H. Taylor, C. M. Maguire, and G. H. Stokes. "Period derivatives and improved parameters for 66 pulsars." *Astrophysical Journal* 332 (1988): 762–69.

Dickel, J. R., and E. W. Greisen. "The evolution of the radio emission from Cas A." *Astronomy and Astrophysics* 75 (1979): 44–53.

Enders, A., and G. Nimtz. "On superluminal barrier traversal." *Physical Review E* 48 (1993): 632.

Erickson, W. C., et al. "Very long baseline interferometer observations of Taurus A and other sources at 121.6 MHz." *Astrophysical Journal* 177 (1972): 101.

Fesen, R. A., and B. Staker. "The structure and motion of the Crab nebula jet." *Monthly Notices of the Royal Astronomical Society* 263 (1993): 69–74.

Filippenko, A. V., and V. Radhakrishnan. "Pulsar nulling and drifting subpulse phase memory." *Astrophysical Journal* 263 (1982): 828–34.

Fowler, L. A., G. A. E. Wright, and D. Morris. "Unusual properties of the pulsar PSR 1822-09." *Astronomy and Astrophysics* 93 (1981): 54–61.

Fruchter, A. S., D. R. Stinebring, and J. H. Taylor. "A millisecond pulsar in an eclipsing binary." *Nature* 333 (1988): 237–39.

Fruchter, A. S., et al. "The eclipsing millisecond pulsar PSR 1957+20." *Astrophysical Journal* 351 (1990): 642–50.

Gold, T. "Rotating neutron stars as the origin of the pulsating radio sources." *Nature* 218 (1968): 731–32.

———. "Apollo II observations of a remarkable glazing phenomenon on the lunar surface." *Science* 165 (1969): 1345–49.

Gower, J., and E. Argyle. "Detection of strong interpulses from NP 0532." *Astrophysical Journal* 171 (1972): L23–L26.

Gull, T. R., and R. A. Fesen. "Deep optical imagery of the Crab Nebula's jet." *Astrophysical Journal* 260 (1982): L75–L78.

Gupta, Y., B. J. Rickett, and A. G. Lyne. "Refractive Interstellar Scintillation in Pulsar Dynamic Spectra." *Monthly Notices of the Royal Astronomical Society* 269 (1994): 1035–68.

Hammer, C. U., H. B. Clausen, and C. C. Langway Jr. "50,000 years of recorded global volcanism." *Climatic Change* 35 (1997): 1–15.

Hankins, T. H. "Microsecond intensity variations in the radio emissions from CP 0950." *Astrophysical Journal* 169 (1971): 487–94.

Hankins, T. H., and J. M. Cordes. "Interpulse emission from pulsar 0950+08: How many poles?" *Astrophysical Journal* 249 (1981): 241.

Hankins, T. H., and D. A. Moffet. "Multifrequency radio observations of the Crab pulsar." *Astrophysical Journal* 468 (1996): 779–83.

Harnden, F. R. "Einstein observations of the Crab nebula pulsar." *Astrophysical Journal* 283 (1984): 279–85.

Hawkes, N. "Going faster than light." *London Times,* April 3, 1995, p. 14.

Heiles, C., D. B. Campbell, and J. M. Rankin. "Pulsar NP 0532: Properties and systematic polarization of individual strong pulses at 430 MHz." *Nature* 226 (1970): 529–31.

Hessels, J. W., et al. "A 20 cm search for pulsars in globular clusters with Arecibo and the GBT." In *Young Neutron Stars and Their Environments,* edited by F. Camilo and B. Gaensler, IAU Symposium, vol. 218, 2004. Available online at arxiv.org/abs/astro-ph/0402182.

Hewish, A., S. Bell, J. Pilkington, P. Scott, and R. Collins. "Observation of a rapidly pulsating radio source." *Nature* 217 (1968): 709–13.

Hill, P. *Unconventional Flying Objects: A Scientific Analysis.* Charlottesville, Va.: Hampton Roads Publishing, 1995.

Hobbs, G., et al. "A statistical study of 233 pulsar proper motions." *Monthly Notices of the Royal Astronomical Society* 360 (2005): 974–92.

Hughen, K. A., et al. "Deglacial changes in ocean circulation from an extended radiocarbon calibration." *Nature* 391 (1998): 65–68.

Huguenin, G. R., J. H. Taylor, and T. H. Troland. "The radio emission from pulsar MP 0031-07." *Astrophysical Journal* 162 (1970): 727–35.

Hyman, S. D., et al. "A new radio detection of the bursting source GCRT J145–3009." 2005, eprint: www.arxiv.org/abs/astro-ph/0508264.

Ilyes. *An Hypothesis: The Transmission of a Crop Circle,* 1996. Online at www.cropcircleconnector.com/ilyes/Abouthy.html.

Intel. "Towards flight without stress or strain . . . or weight." *Interavia* 11, no. 5 (1956): 373–74.

Isaacman, R. "NP 0532 and a hole in the Crab Nebula." *Nature* 268 (1977): 317–18.

Ishii, T. K., and G. C. Giakos. "Radio Messages Faster Than Light." *Microwaves & RF* 30 (August 1991): 114–19.

———. "Rapid Pulsed Microwave Propagation." *IEEE Microwave & Guided Wave Letters* 1, no. 12 (December 1991): 374–75

Jastrow, R., and M. H. Thompson. *Astronomy: Fundamentals and Frontiers.* New York: John Wiley & Sons, 1977.

Johnston, S., et al. "Discovery of a very bright, nearby binary millisecond pulsar." *Nature* 361 (1995): 613–15.

————. "High time-resolution observations of the Vela pulsar." *Astrophysical Journal* 549 (2001): L149.

Jouzel, J., et al. "Vostok ice core: A continuous isotope temperature record over the last climatic cycle (160,000 years)." *Nature* 329 (1987): 403–408.

Kanbach, G., et al. "Detailed characteristics of the high-energy gamma radiation from PSR 0833-45 measured by COS-B." *Astronomy and Astrophysics* 90 (1980): 163–69.

Kramer, M., S. Johnston, and W. van Straten, "High-resolution single-pulse studies of the Vela pulsar," *Monthly Notices of the Royal Astronomical Society* 334 (2002): 523.

Kulkarni, S. R., and J. J. Hester. "Discovery of a nebula around PSR 1957+20." *Nature* 335 (1988): 801–804.

LaViolette, P. A. "Galactic explosions, cosmic dust invasions, and climatic change." Ph.D. dissertation. Portland, Ore.: Portland State University, 1983.

————. "An introduction to subquantum kinetics." *International Journal of General Systems* 11 (1985): 281–345.

————. "Cosmic-ray volleys from the Galactic center and their recent impact on the Earth environment." *Earth, Moon, and Planets* 37 (1987): 241–86.

————. "A Tesla wave physics for a free energy universe." *Proceedings of the 1990 International Tesla Symposium*, 5.1–5.19. Colorado Springs: International Tesla Society, 1991.

————. "The planetary-stellar mass-luminosity relation: Possible evidence of energy nonconservation?" *Physics Essays* 5, no. 4 (1992): 536–44.

————. "The U.S. Antigravity Squadron." In *Electrogravitics Systems: Reports on a New Propulsion Methodology*, edited by T. Valone. Washington, D.C.: Integrity Research Institute, 1994.

————. "Evidence that radio pulsars may be artificial beacons of ETI origin." 195th American Astronomical Society meeting, Atlanta, Georgia, January 2000.

————. *Subquantum Kinetics: A Systems Approach to Physics and Cosmology.* Niskayuna, N.Y.: Starlane Publications, 2003.

————. *Genesis of the Cosmos: The Ancient Science of Continuous Creation.* Rochester, Vt.: Bear & Co., 2004.

————. *Earth Under Fire: Humanity's Survival of the Ice Age.* Rochester, Vt.: Bear & Co., 2005.

————. "Solar cycle variations in ice acidity at the end of the last ice age: Possible marker of a climatically significant interstellar dust incursion." *Planetary & Space Science* 53, no. 4 (2005): 385–93. Eprint: arXiv.org/abs/physics/0502019.

————. "Evidence for a global warming at the Termination I boundary and its possible extraterrestrial cause." 2005. Eprint: arXiv.org/abs/physics/0503158.

————. *Galactic Superwaves and Their Impact on the Earth*. 2005 CD-ROM update of "Galactic Explosions, Cosmic Dust Invasions, and Climatic Change."

Levengood, W. C. "Anatomical Anomalies in Crop Formation Plants." *Physiologia Plantarum* 92 (1994): 356–63.

Lorenzen, C., and J. Lorenzen. *UFO: The Whole Story*. New York: Signet, 1968.

Lyne, A. G., and F. Graham-Smith. *Pulsar Astronomy*. London: Cambridge University Press, 1998.

Lyne, A. G., R. S. Pritchard, F. Graham-Smith, and F. Camilo. "Very low braking index for the Vela pulsar." *Nature* 381 (1996): 497–98.

Manchester, R. N., J. H. Taylor, and G. R. Huguenin. "Observations of pulsar radio emission. II. Polarization of individual pulses." *Astrophysical Journal* 196 (1975): 83–112.

Manchester, R. N., and J. H. Taylor. "Recent Observations of Pulsars." *Annual Reviews of Astronomy and Astrophysics* 15 (1977): 19–44.

————. "Observed and derived parameters for 330 pulsars." *Astronomical Journal* 86 (1981): 1953–73.

Mason, H. "Bright Skies: Top-Secret Weapons Testing?" *Nexus,* April–May 1997: 41–47, 78.

McCulloch, P. M., P. A. Hamilton, G. W. R. Royle, and R. N. Manchester. "Daily observations of a large period jump of the Vela pulsar." *Nature* 302 (1983): 319–21.

McMoneagle, J. *The Ultimate Time Machine*. Charlottesville, Va.: Hampton Roads, 1999.

Michanowsky, G. *The Once and Future Star*. New York: Barnes and Noble, 1979.

Misner, C., K. Thorne, and J. A. Wheeler. *Gravitation*. San Francisco: Freeman and Co., 1973.

Moffet, D. A., and T. H. Hankins. "Multifrequency radio observations of the Crab pulsar." *Astrophysical Journal* 468 (1996): 779–83.

Pepper, D. M. "Applications of optical phase conjugation." *Scientific American* 254 (January 1986): 74–83.

Plutarch. "Lucullus." In *Plutarch's Lives,* vol. 2, translated by B. Perrin, 495–97. Cambridge, Mass.: Harvard University Press, 1968.

Podkletnov, E., and G. Modanese. "Impulse gravity generator based on charged YBa2Cu3O7-y superconductor with composite crystal structure." August 2001. Eprint: arXiv.org/abs/physics/0108005.

———. "Investigation of high voltage discharges in low pressure gases through large ceramic superconducting electrodes." *Journal of Low Temperature Physics* 132 (2003): 239–59. Eprint: arXiv.org/abs/physics/0209051.

Preston, R. "The eclipsing death star." *Discovery,* August 1988: 41–46.

Raisbeck, G. M., et al. "^{10}Be deposition at Vostok, Antarctica, during the last 50,000 years and its relationship to possible cosmogenic production variations during this period." In *The Last Deglaciation: Absolute and Radiocarbon Chronologies* (NATO ASI Series, vol. 12), edited by E. Bard and W. Broecker, 127–39. Heidelberg: Springer-Verlag, 1992.

Reynolds, S. P., and D. C. Ellison. "Electron acceleration in Tycho's and Kepler's supernova remnants: Spectral evidence of Fermi shock acceleration." *Astrophysical Journal* 399 (1992): L75–L78.

Schmidt, G. D., J. Angel, and E. Beaver. "The small-scale polarization of the Crab Nebula." *Astrophysical Journal* 227 (1979): 106–13.

Sharp, N. A. "Millisecond time resolution with the Kitt Peak Photoncounting array." *Publications of the Astronomical Society of the Pacific* 104 (1992): 263–69.

Shkunov, V. V., and B. Y. Zel'dovich. "Optical phase conjugation." *Scientific American* 253 (December 1985): 54–59.

Silva, F. "Music in the Fields." *Atlantis Rising,* no. 14 (1998): 42–43.

Steiger, B. *Alien Meetings.* New York: Ace, 1978.

Sturrock, P. A., et al. "Physical evidence related to UFO reports." *Journal of Scientific Exploration* 12 (1998): 179–229.

Sullivan, W. *Black Holes.* Garden City, N.Y.: Anchor Press, 1979.

Taylor, J. H., and R. N. Manchester. "Recent Observations of Pulsars." *Annual Reviews of Astronomy and Astrophysics* 15 (1977): 19–44.

Taylor, J. H., R. N. Manchester, and G. R. Huguenin. "Observations of pulsar radio emission. I. Total-intensity measurements of individual pulses." *Astrophysical Journal* 195 (1975): 513–28.

Taylor, J. H., R. N. Manchester, and A. G. Lyne. "Catalog of 558 pulsars." *Astrophysical Journal Supplement Series* 88 (1993): 529–68.

Taylor, J. H., and D. R. Stinebring. "Recent progress in the understanding of pulsars." *Annual Reviews of Astronomy and Astrophysics* 24 (1986): 285–327.

Thompson, R. L. *Alien Identities: Ancient Insights into Modern UFO Phenomena.* San Diego, Calif.: Govardhan Hill Publishing, 1993.

Trimble, V. "Motions and Structure of the Filamentary Envelope of the Crab Nebula." *Astronomical Journal* 73 (1968): 535–47.

Tucker, W. H. "Supernova in the sail." *Star and Sky* 2, no. 1 (1980): 36.

Vassilatos, G. *Secrets of Cold War Technology: Project HAARP and Beyond.* Bayside, Calif.: Borderland Sciences, 1996.

Velusamy, T. "Radio detection of a jet in the Crab Nebula." *Nature* 308 (1984): 251–52.

Wallace, H. W. "Method and apparatus for generating a secondary gravitational force field." U.S. patent no. 3,626,605, December 14, 1971.

———. "Method and apparatus for generating a dynamic force field." U.S. patent no. 3,626,606, Dec. 14, 1971.

Webre, A. L. *Exopolitics: Politics, Government, and Law in the Universe.* Vancouver, B.C.: Universe Books, 2005.

West, John Anthony. *Serpent in the Sky: The High Wisdom of Ancient Egypt.* Wheaton, Ill.: Quest, 1993.

Wilson, R. B., and G. J. Fishman. "The pulse profile of the Crab pulsar in the energy range 45 keV-1.2 MeV." *Astrophysical Journal* 269 (1983): 273–80.

Winn, J. "The life of a neutron star." *Sky & Telescope,* July 1999: 34.

Wolszczan, A., and D. A. Frail. "A planetary system around the millisecond pulsar PSR 1257+12." *Nature* 355 (1992): 145–47.

Woodward, J. "A new experimental approach to Mach's principle and relativistic gravitation." *Foundations of Physics Letters,* vol. 3, no. 5 (1990).

Wright, G. A. E., and L. A. Fowler. "Mode-changing and quantized subpulse drift-rates in pulsar PSR 2319+60." *Astronomy and Astrophysics* 101 (1981): 356–61.

Zeitlin, G. "Are pulsar signals evidence of astro-engineered signalling systems?" *New Frontiers in Science* 1 (Summer 2002).

Zook, H. A., J. B. Hartung, and D. Storzer. "Solar flare activity: Evidence for large-scale changes in the past." *Icarus* 32 (1977): 106–26.

INDEX

BOOKS OF RELATED INTEREST

GENESIS OF THE COSMOS
The Ancient Science of Continuous Creation
by Paul A. LaViolette, Ph.D.

EARTH UNDER FIRE
Humanity's Survival of the Ice Age
by Paul A. LaViolette, Ph.D.

SCIENCE AND THE AKASHIC FIELD
An Integral Theory of Everything
by Ervin Laszlo

SCIENCE AND THE REENCHANTMENT OF THE COSMOS
The Rise of the Integral Vision of Reality
by Ervin Laszlo

MATRIX OF CREATION
Sacred Geometry in the Realm of the Planets
by Richard Heath

CHAOS, CREATIVITY, AND COSMIC CONSCIOUSNESS
by Rupert Sheldrake, Terence McKenna,
and Ralph Abraham

ALTERNATIVE SCIENCE
Challenging the Myths of the Scientific Establishment
by Richard Milton

SHATTERING THE MYTHS OF DARWINISM
by Richard Milton

Inner Traditions • Bear & Company
P.O. Box 388
Rochester, VT 05767
1-800-246-8648
www.InnerTraditions.com

Or contact your local bookseller